Air Quality Guidelines
for Europe

Second Edition

WHO Library Cataloguing in Publication Data

Air quality guidelines for Europe ; second edition

(WHO regional publications. European series ; No. 91)

1.Air pollution – prevention and control
2.Air pollutants – adverse effects – toxicity
3.Air pollution, Indoor – prevention and control
4.Environmental exposure 5.Guidelines 6.Europe I.Series

ISBN 92 890 1358 3 (NLM Classification: WA 754)
ISSN 0378-2255

Text editing: Frank Theakston
Cover design: G. Gudmundsson

Air Quality Guidelines

for Europe

Second Edition

WHO Regional Publications, European Series, No. 91

ISBN 1358 3
ISSN 0378-2255

The designations employed and the presentation of the material in this publication do not imply the expression of any opinion whatsoever on the part of the Secretariat of the World Health Organization concerning the legal status of any country, territory, city or area or of its authorities, or concerning the delimitation of its frontiers or boundaries. The names of countries or areas used in this publication are those that obtained at the time the original language edition of the book was prepared.

The mention of specific companies or of certain manufacturers' products does not imply that they are endorsed or recommended by the World Health Organization in preference to others of a similar nature that are not mentioned. Errors and omissions excepted, the names of proprietary products are distinguished by initial capital letters.

The views expressed in this publication are those of the participants in the meetings and do not necessarily represent the decisions or the stated policy of the World Health Organization.

Contents

Foreword

Clean air is considered to be a basic requirement for human health and wellbeing. In spite of the introduction of cleaner technologies in industry, energy production and transport, air pollution remains a major health risk. Recent epidemiological studies have provided evidence that in Europe hundreds of thousands of premature deaths are attributed to air pollution. The World Health Organization has been concerned with air pollution and its impact on human health for more than 40 years. In 1987 these activities culminated in the publication of the first edition of Air quality guidelines for Europe. *It was the aim of the guidelines to provide a basis for protecting public health from adverse effects of air pollutants, to eliminate or reduce exposure to hazardous air pollutants, and to guide national and local authorities in their risk management decisions. The guidelines were received with great enthusiasm and found wide application in environmental decision-making in the European Region as well as in other parts of the world.*

Since the publication of the first edition, new scientific data in the field of air pollution toxicology and epidemiology have emerged and new developments in risk assessment methodology have taken place. It was therefore necessary to update and revise the existing guidelines. Starting in 1993, the Bilthoven Division of the WHO European Centre for Environment and Health undertook this process in close cooperation with WHO headquarters and the European Commission. More than 100 experts contributed to the preparation of the background documents or participated in the scientific discussions that led to the derivation of guideline values for a great number of air pollutants. WHO is most grateful for their contribution and expert advice. Financial support received from the European Commission, the Swedish Environmental Protection Agency and the Government of the Netherlands during the preparation of the second edition of the guidelines made this effort possible and is warmly acknowledged.

The guidelines are a contribution to HEALTH21, *the health for all policy framework for the WHO European Region. This states that, by the year 2015, people in the Region should live in a safer physical environment, with exposure to contaminants hazardous to health at levels not exceeding internationally agreed standards. WHO is therefore pleased to see that the revised air quality guidelines are being used as a starting point for the derivation of legally binding limit values in the framework of the EU Air Quality Directive. Also, the global guidelines for air quality, recently issued by WHO headquarters, are based on the revised guidelines for Europe.*

Thus, the work and efforts of everybody who contributed to the revision of the guidelines has already had an important impact. It is expected that the publication of this second edition will provide the Member States with a sound basis for improving human health by ensuring adequate air quality for all. I should like to warmly thank all the WHO staff who made this important endeavour possible.

Marc A. Danzon
WHO Regional Director for Europe

Preface

The first edition of the WHO *Air quality guidelines for Europe* was published in 1987. Since then new data have emerged and new developments in risk assessment methodology have taken place, necessitating the updating and revision of the existing guidelines. The Bilthoven Division of the WHO European Centre for Environment and Health has undertaken this process in close cooperation with the International Programme on Chemical Safety (IPCS) and the European Commission.

At the start of the process, the methods to be used in the risk assessment process, the use of the threshold concept, the application of uncertainty factors, and the quantitative risk assessment of carcinogens were discussed, and the approach to be used was agreed on. In setting priorities for the compounds to be reviewed, a number of criteria were established: (*a*) the compound (or mixture) posed a widespread problem in terms of exposure sources; (*b*) the potential for personal exposure was large; (*c*) new data on health or environmental impact had emerged; (*d*) monitoring had become feasible since the previous evaluation; and (*e*) a positive trend in ambient air concentrations was evident. Application of these criteria has resulted in the selection of the air pollutants addressed in the review process.

It is the aim of the guidelines to provide a basis for protecting public health from adverse effects of air pollutants and to eliminate or reduce exposure to those pollutants that are known or likely to be hazardous to human health or wellbeing. The guidelines are intended to provide background information and guidance to (inter)national and local authorities in making risk assessment and risk management decisions. In establishing pollutant levels below which exposure – for life or for a given period of time – does not constitute a significant public health risk, the guidelines provide a basis for setting standards or limit values for air pollutants.

Although the guidelines are considered to be protective to human health they are by no means a "green light" for pollution, and it should be stressed that attempts should be made to keep air pollution levels as low as practically achievable. In addition, it should be noted that in general the guidelines do not differentiate between indoor and outdoor air exposure because, although the site of exposure determines the composition of the air and the concentration of the various pollutants, it does not directly affect the exposure–response relationship.

In general, the guidelines address single pollutants, whereas in real life exposure to mixtures of chemicals occurs, with additive, synergistic or antagonistic effects. In dealing with practical situations or standard-setting procedures, therefore, consideration should be given to the interrelationships between the various air pollutants. It should be emphasized, however, that the guidelines are health-based or based on environmental effects, and are not standards *per se*. In setting legally binding standards, considerations such as prevailing exposure levels, technical feasibility, source control measures, abatement strategies, and social, economic and cultural conditions should be taken into account.

It is a policy issue to decide which specific groups at risk should be protected by the standards and what degree of risk is considered to be acceptable. These decisions are influenced by differences in risk perception among the general population and the various stakeholders in the process, but also by differences in social situations in different countries, and by the way the risks associated with air pollution are compared with risks from other environmental exposures or human activities. National standards may therefore differ from country to country and may be above or below the respective WHO guideline value.

This publication includes an introduction on the nature of the guidelines and the methodology used to establish guideline values for a number of air pollutants. In addition, it describes the various aspects that need to be considered by national or local authorities when guidelines are transformed into legally binding standards. For the pollutants addressed, the sections on "Health risk evaluation" and "Guidelines" describe the most relevant considerations that have led to the recommended guideline values. For detailed information on exposure and on the potential health effects of the reviewed pollutants, the reader is referred to the Regional Office's web site, where the background documents on the individual air pollutants can be accessed.

F.X. Rolaf van Leeuwen and Michal Krzyzanowski
WHO European Centre for Environment and Health
Bilthoven, Netherlands

PART I

GENERAL

CHAPTER I

Introduction

Human beings need a regular supply of food and water and an essentially continuous supply of air. The requirements for air and water are relatively constant (10–20 m^3 and 1–2 litres per day, respectively). That all people should have free access to air and water of acceptable quality is a fundamental human right. Recognizing the need of humans for clean air, in 1987 the WHO Regional Office for Europe published *Air quality guidelines for Europe (1)*, containing health risk assessments of 28 chemical air contaminants.

These guidelines can be seen as a contribution to target 10 of HEALTH21, the health for all policy framework for the WHO European Region as formulated in 1999 *(2)*. This target states that by the year 2015, people in the Region should live in a safer physical environment, with exposure to contaminants hazardous to health at levels not exceeding internationally agreed standards. The achievement of this target will require the introduction of effective legislative, administrative and technical measures for the surveillance and control of both outdoor and indoor air pollution, in order to comply with criteria to safeguard human health. Unfortunately, this ambitious objective is not likely to be met in the next few years in many areas of Europe. Improvement in epidemiological research over the 1990s and greater sensitivity of the present studies have revealed that people's health may be affected by exposures to much lower levels of some common air pollutants than believed even a few years ago. While the no-risk situation is not likely to be achieved, a minimization of the risk should be the objective of air quality management, and this is probably a major conceptual development of the last few years.

Various chemicals are emitted into the air from both natural and man-made (anthropogenic) sources. The quantities may range from hundreds to millions of tonnes annually. Natural air pollution stems from various biotic and abiotic sources such as plants, radiological decomposition, forest fires, volcanoes and other geothermal sources, and emissions from land and water. These result in a natural background concentration that varies according to local sources or specific weather conditions. Anthropogenic air pollution has existed at least since people learned to use fire, but it has increased rapidly since industrialization began. The increase in air pollution resulting from the expanding use of fossil energy sources and the growth in the manufacture and use of chemicals has been accompanied by mounting public awareness of and concern about its detrimental effects on health and the environment. Moreover, knowledge of the nature, quantity, physico-chemical behaviour and effects of air pollutants has greatly increased in recent years. Nevertheless, more needs to be known. Certain aspects of the health effects of air pollutants require further assessment; these include newer scientific areas such as developmental toxicity. The proposed guideline values will undoubtedly be changed as future studies lead to new information.

The impact of air pollution is broad. In humans, the pulmonary deposition and absorption of inhaled chemicals can have direct consequences for health. Nevertheless, public health can also be indirectly affected by deposition of air pollutants in environmental media and uptake by plants and animals, resulting in chemicals entering the food chain or being present in drinking-water and thereby constituting additional sources of human exposure. Furthermore, the direct effects of air pollutants on plants, animals and soil can influence the structure and function of ecosystems, including their self-regulation ability, thereby affecting the quality of life.

In recent decades, major efforts have been made to reduce air pollution in the European Region. The emission of the main air pollutants has declined significantly. The most pronounced effect is observed for sulfur dioxide: its total emission was reduced by about 50% in the period 1980–1995. Reduction of emission of nitrogen oxides was smaller and was observed only after 1990: total emission declined by about 15% in the period from 1990 to 1995 *(3)*. The reduction of sulfur dioxide emission is reflected by declining concentrations in ambient air in urban areas. Trends in concentrations of other pollutants in urban air, such as nitrogen dioxide or particulate matter, are less clear and it is envisaged that these pollutants still constitute a risk to human health *(4)*.

Many countries of the European Region encounter similar air pollution problems, partly because pollution sources are similar, and in any case air pollution does not respect national frontiers. The subject of the transboundary long-range transport of air pollution has received increasing attention in Europe over the last decade. International efforts to combat emissions are undertaken, for instance within the framework of the Convention on Long-range Transboundary Air Pollution established by the United Nations Economic Commission for Europe *(5, 6)*.

The task of reducing levels of exposure to air pollutants is a complex one. It begins with an analysis to determine which chemicals are present in the air, at what levels, and whether likely levels of exposure are hazardous to human health and the environment. It must then be decided whether an unacceptable risk is present. When a problem is identified, mitigation strategies should be developed and implemented so as to prevent excessive risk to public health in the most efficient and cost-effective way.

Analyses of air pollution problems are exceedingly complicated. Some are national in scope (such as the definition of actual levels of exposure of the population, the determination of acceptable risk, and the identification of the most efficient control strategies), while others are of a more basic character and are applicable in all countries (such as analysis of the relationships between chemical exposure levels, and doses and their effects). The latter form the basis of these guidelines.

The most direct and important source of air pollution affecting the health of many people is tobacco smoke. Even those who do not smoke may inhale the smoke produced by others ("passive smoking"). Indoor pollution in general and occupational exposure in particular also contribute substantially to overall human exposure: indoor concentrations of nitrogen dioxide, carbon monoxide, respirable particles, formaldehyde and radon are often higher than outdoor concentrations *(7)*.

Outdoor air pollution can originate from a single point source, which may affect only a relatively small area. More often, outdoor air pollution is caused by a mixture of pollutants from a variety of diffuse sources, such as traffic and heating, and from point sources. Finally, in addition to those emitted by local sources, pollutants transported over medium and long distances contribute further to the overall level of air pollution.

The relative contribution of emission sources to human exposure to air pollution may vary according to regional and lifestyle factors. Although, as far as some pollutants are concerned, indoor air pollution will be of greater importance than outdoor pollution, this does not diminish the importance of outdoor pollution. In terms of the amounts of substances released, the latter is far more important and may have deleterious effects on animals, plants and materials as well as adverse effects on human health. Pollutants produced outdoors may penetrate into the indoor environment and may affect human health by exposure both indoors and outdoors.

NATURE OF THE GUIDELINES

The primary aim of these guidelines is to provide a basis for protecting public health from adverse effects of air pollution and for eliminating, or reducing to a minimum, those contaminants of air that are known or likely to be hazardous to human health and wellbeing. In the present context, guidelines are not restricted to a numerical value below which exposure for a given period of time does not constitute a significant health risk; they also include any kind of recommendation or guidance in the relevant field.

The guidelines are intended to provide background information and guidance to governments in making risk management decisions, particularly in setting standards, but their use is not restricted to this. They also provide information for all who deal with air pollution. The guidelines may be used in planning processes and various kinds of management decisions at community or regional level.

When guideline values are indicated, this does not necessarily mean that they should be used as the starting point for producing general countrywide standards, monitored by a comprehensive network of control stations. In the case of some pollutants, guideline values may be of use mainly for carrying out local control measures around point sources. To aid in this process, information on major sources of pollutants has been provided.

It should be emphasized that when numerical air quality guideline values are given, these values are not standards in themselves. Before transforming them into legally binding standards, the guideline values must be considered in the context of prevailing exposure levels, technical feasibility, source control measures, abatement strategies, and social, economic and cultural conditions (see Chapter 4). In certain circumstances there may be valid reasons to pursue policies that will result in pollutant concentrations above or below the guideline values.

Although these guidelines are considered to protect human health, they are by no means a "green light" for pollution. It should be stressed that attempts should be made to keep air pollution levels as low as practically achievable.

Ambient air pollutants can cause a range of significant effects that require attention: irritation, odour annoyance, and acute and long-term toxic effects. Numerical air quality guidelines either indicate levels combined with exposure times at which no adverse effect is expected in terms of noncarcinogenic endpoints, or they provide an estimate of lifetime cancer risk arising from those substances that are proven human carcinogens or carcinogens with at least limited evidence of human carcinogenicity. It should be noted that the risk estimates for carcinogens do not indicate a safe level, but they are presented so that the carcinogenic potencies of different carcinogens can be compared and an assessment of overall risk made.

It is believed that inhalation of an air pollutant in concentrations and for exposure times below a guideline value will not have adverse effects on health and, in the case of odorous compounds, will not create a nuisance of indirect health significance. This is in line with the definition of health: a state of complete physical, mental and social wellbeing and not merely the absence of disease or infirmity *(8)*. Nevertheless, compliance with recommendations regarding guideline values does not guarantee the absolute exclusion of effects at levels below such values. For example, highly sensitive groups such as those impaired by concurrent disease or other physiological limitations may be affected at or near concentrations referred to in the guideline values. Health effects at or below guideline values may also result from combined exposure to various chemicals or from exposure to the same chemical by multiple routes.

It is important to note that guidelines have been established for single chemicals. Mixtures of chemicals can have additive, synergistic or antagonistic effects. In general, our knowledge of these interactions is rudimentary. One exception can be found in a WHO publication on summer and winter smog *(9)*, which deals with commonly recurring mixtures of air pollutants.

In preparing this second edition of the guidelines, emphasis has been placed on providing data on the exposure–response relationships of the pollutants considered. It is expected that this will provide a basis for estimating the risk to health posed by monitored concentrations of these pollutants.

Although health effects were the major consideration in establishing the guidelines, evidence of the effects of pollutants on terrestrial vegetation was also considered and guideline values were recommended for a few substances (see Part III). These ecological guidelines have been established because, in the long term, only a healthy total environment can guarantee human health and wellbeing. Ecological effects on life-forms other than humans and plants have not been discussed since they are outside the scope of this book.

The guidelines do not differentiate between indoor and outdoor exposure (with the exception of exposure to mercury) because, although the sites of exposure influence the type and concentration of air pollutants, they do not directly affect the basic exposure–effect relationships. Occupational exposure has been considered in the evaluation process, but it was not a main focus of attention as these guidelines relate to the general population. However, it should be noted that occupational exposure may add to the effects of environmental exposure. The guidelines do not apply to very high short-term concentrations that may result from accidents or natural disasters.

The health effects of tobacco smoking have not been assessed here, the carcinogenic effects of smoking having already been evaluated by IARC in 1986 *(10)*. Neither have the effects of air pollutants on climate been considered, since too many uncertainties remain to allow a satisfactory evaluation of possible adverse health and environmental effects. Possible changes of climate, however, should be investigated very seriously by the appropriate bodies because their overall consequences, for example the "greenhouse effect", may go beyond direct adverse effects on human health or ecosystems.

PROCEDURES USED IN THE UPDATING AND REVISION PROCESS

The first step in the process of updating and revising the guidelines was the selection of pollutants. Air pollutants of special environmental and health significance to countries of the European Region were identified and selected by a WHO planning group in 1993 *(11)* on the basis of the following criteria:

(*a*) whether substances or mixtures posed a widespread problem in terms of sources;
(*b*) the ubiquity and abundance of the pollutants where the potential for exposure was large, taking account of both outdoor and indoor exposure;
(*c*) whether significant new information on health effects had become available since the publication of the first edition of the guidelines;
(*d*) the feasibility of monitoring;
(*e*) whether significant non-health (e.g. ecotoxic) effects could occur; and
(*f*) whether a positive trend in ambient levels was evident.

During the deliberations of the planning group, compounds that had not been dealt with in the first edition of the guidelines were also considered, including butadiene, fluoride, compounds associated with global warming and with alterations in global air pollution (and possibly with secondary health effects), and compounds associated with the development of alternative fuels and new fuel additives. Other factors affecting selection included the timetable of the project, and the fact that only those substances for which sufficient documentation was available could be considered.

The existence of relevant WHO Environmental Health Criteria documents was of great value in this respect. On the basis of these considerations, the following 35 pollutants were selected to be included in this second edition of the guidelines:

Organic air pollutants

Acrylonitrile[1]
Benzene
Butadiene
Carbon disulfide[1]
Carbon monoxide
1,2-Dichloroethane[1]
Dichloromethane
Formaldehyde
Polycyclic aromatic hydrocarbons (PAHs)
Polychlorinated biphenyls (PCBs)
Polychlorinated dibenzodioxins and dibenzofurans (PCDDs/PCDFs)
Styrene
Tetrachloroethylene
Toluene
Trichloroethylene
Vinyl chloride[1]

Inorganic air pollutants

Arsenic
Asbestos[1]
Cadmium
Chromium
Fluoride
Hydrogen sulfide[1]
Lead
Manganese
Mercury
Nickel
Platinum
Vanadium[1]

Classical air pollutants

Nitrogen dioxide
Ozone and other photochemical oxidants
Particulate matter
Sulfur dioxide

Indoor air pollutants

Environmental tobacco smoke
Man-made vitreous fibres
Radon

[1] 1987 evaluation retained, not re-evaluated.

In addition to the 35 pollutants listed above, this second edition expands on the ecological effects presented in the first edition in an enlarged section examining the ecotoxic effects of sulfur dioxide (including sulfur and total acid deposition), nitrogen dioxide (and other nitrogen compounds, including ammonia) and ozone.

To carry out the evaluation process, the planning group established a number of working groups on:

- methodology and format
- ecotoxic effects
- classical air pollutants
- inorganic air pollutants
- certain indoor air pollutants
- polychlorinated biphenyls, dioxins and furans
- volatile organic pollutants.

The dates of the meetings of these working groups and the membership are listed in Annex I.

Before the meeting of each working group, scientific background documents providing in-depth reviews of each pollutant were prepared as a basis for discussion. Guidelines were established on the basis of these discussions. After each meeting, a text on each pollutant or pollutant group was drafted on the basis of the amended background documents, incorporating the working group's conclusions and recommendations. The draft report of the working group was then circulated to all participants for their comments and corrections. A final consultation group was then convened to critically review the documents for clarity of presentation, adequacy of description of the rationale supporting each guideline and consistency in the application of criteria, and with a view to possibly considering newly emerged information. The process concluded with a review of the recommendations and conclusions of all the working groups.

It was appreciated, during preparation of this second edition, that the expanded range of pollutants being considered and the considerably expanded database available for some pollutants would lead to a significant lengthening of the text. It was therefore decided to publish in this volume summaries of the data on which the guidelines are based. The full background evaluation will become progressively available on the Regional Office's web site.

As in the first edition, detailed referencing of the relevant literature has been provided with indications of the periods covered by the reviews of individual pollutants. Every effort has been made to ensure that the material provided is as up-to-date as possible, although the extended period of preparation of this second edition has inevitably meant that some sections refer to more recently published material than others.

During the preparation of the second edition, the Directorate-General for Environment, Nuclear Safety and Civil Protection (DGXI) of the European Commission developed a Framework Directive and a number of daughter directives dealing with individual pollutants. It was agreed with the Commission that the final drafts of the revised WHO guideline documents would provide a starting point for discussions by the Commission's working groups aiming at setting legally binding limit values for air quality in the European Union.

REFERENCES

1. *Air quality guidelines for Europe*. Copenhagen, WHO Regional Office for Europe, 1987 (WHO Regional Publications, European Series, No. 23).
2. *HEALTH21. The health for all policy framework for the WHO European Region*. Copenhagen, WHO Regional Office for Europe, 1999 (European Health for All Series, No. 6).
3. *Europe's environment: the second assessment*. Copenhagen, European Environment Agency, 1998.
4. *Overview of the environment and health in Europe in the 1990s: Third Ministerial Conference on Environment and Health, London, 16–18 June 1999*. Copenhagen, WHO Regional Office for Europe, 1998 (document EUR/ICP/EHCO 02 02 05/6).
5. *Convention on Long-range Transboundary Air Pollution. Strategies and policies for air pollution abatement. 1994 major review*. New York, United Nations, 1995 (ECE/EB.AIR/44).
6. *Convention on Long-range Transboundary Air Pollution. Major review of strategies and policies for air pollution abatement*. New York, United Nations, 1998 (EB.AIR/1998/3, Add.1).
7. WHO EUROPEAN CENTRE FOR ENVIRONMENT AND HEALTH. *Concern for Europe's tomorrow. Health and the environment in the WHO European Region*. Stuttgart, Wissenschaftliche Verlagsgesellschaft, 1995.
8. *Constitution of the World Health Organization*. Geneva, World Health Organization, 1985.

9. *Acute effects on health of smog episodes.* Copenhagen, WHO Regional Office for Europe, 1992 (WHO Regional Publications, European Series, No. 43).
10. *Tobacco smoking.* Lyons, International Agency for Research on Cancer, 1986 (IARC Monographs on the Evaluation of the Carcinogenic Risk of Chemicals to Humans, Vol. 38).
11. *Update and revision of the air quality guidelines for Europe. Report of a WHO planning meeting.* Copenhagen, WHO Regional Office for Europe, 1994 (document EUR/ICP/CEH 230).

Criteria used in establishing guideline values

Relevant information on the pollutants was carefully considered during the process of establishing guideline values. Ideally, guideline values should represent concentrations of chemical compounds in air that would not pose any hazard to the human population. Realistic assessment of human health hazards, however, necessitates a distinction between absolute safety and acceptable risk. To produce a guideline with a high probability of offering absolute safety, one would need a detailed knowledge of dose–response relationships in individuals in relation to all sources of exposure, the types of toxic effect elicited by specific pollutants or their mixtures, the existence or nonexistence of "thresholds" for specified toxic effects, the significance of interactions, and the variation in sensitivity and exposure levels within the human population. Such comprehensive and conclusive data on environmental contaminants are generally unavailable. Very often the relevant data are scarce and the quantitative relationships uncertain. Scientific judgement and consensus therefore play an important role in establishing guidance that can be used to indicate acceptable levels of population exposure. Value judgements are needed and the use of subjective terms such as "adverse effects" and "sufficient evidence" is unavoidable.

Although it may be accepted that a certain risk can be tolerated, the risks to individuals within a population may not be equally distributed: there may be subpopulations that are at considerably increased risk. Therefore, groups at special risk in the general population must be taken specifically into account in the risk management process. Even if knowledge about groups with specific sensitivity is available, unknown factors may exist that change the risk in an unpredictable manner. During the preparation of this second edition of the guidelines, attention has been paid to defining specific sensitive subgroups in the population.

INFORMATION COMMON TO CARCINOGENS AND NONCARCINOGENS

Sources, levels and routes of exposure

Available data are provided on the current levels of human exposure to pollutants from all sources, including the air. Special attention is given to atmospheric concentrations in urban and unpolluted rural areas and in the indoor environment. Where appropriate, concentrations in the workplace are also indicated for comparison with environmental levels. To provide information on the contribution from air in relation to all other sources, data on uptake by inhalation, ingestion from water and food, and dermal contact are given where relevant. For most chemicals, however, data on total human exposure are incomplete.

Toxicokinetics

Available data on toxicokinetics (absorption, distribution, metabolism and excretion) of air pollutants in humans and experimental animals are provided for comparison between test species and humans and for interspecies and intraspecies extrapolation, especially to assess the magnitude of body burden from long-term, low-level exposures and to characterize better the mode of toxic action. Data concerning the distribution of a compound in the body are important in determining the molecular or tissue dose to target organs. It has been appreciated that high-to-low-dose and interspecies extrapolations are more easily carried out using equivalent tissue doses. Metabolites are mentioned, particularly if they are known or believed to exert a greater toxic potential than the parent compound. Additional data of interest in determining the fate of a compound in a living organism include the rate of excretion and the biological half-life. These toxicokinetic parameters should be compared between test species and humans for derivation of interspecies factors where this is possible.

Terminology

The following terms and definitions are taken largely from Environmental Health Criteria No. 170, 1994 *(1)*.

Adverse effect Change in morphology, physiology, growth, development or life span of an organism which results in impairment of functional capacity or impairment of capacity to compensate for additional stress or increase in susceptibility to the harmful effects of other environmental influences.

Benchmark dose (**BMD**) The lower confidence limit of the dose calculated to be associated with a given incidence (e.g. 5% or 10% incidence) of effect estimated from all toxicity data on that effect within that study *(2)*.

Critical effect(s) The adverse effect(s) judged to be most appropriate for the health risk evaluation.

Lowest-observed-adverse-effect level (LOAEL) Lowest concentration or amount of a substance, found by experiment or observation, which causes an adverse alteration of morphology, functional capacity, growth, development or life span of the target organism distinguishable from normal (control) organisms of the same species and strain under the same defined conditions of exposure.

No-observed-adverse-effect level (NOAEL) Greatest concentration or amount of a substance, found by experiment or observation, which causes no detectable adverse alteration of morphology, functional capacity, growth, development or life span of the target organism under defined conditions of exposure. Alterations of morphology, functional capacity, growth, development or life span of the target may be detected which are judged not to be adverse.

Toxicodynamics The process of interaction of chemical substances with target sites and the subsequent reactions leading to adverse effects.

Toxicokinetics The process of the uptake of potentially toxic substances by the body, the biotransformation they undergo, the distribution of the substances and their metabolites in the tissues, and the elimination of the substances and their metabolites from the body. Both the amounts and the concentrations of the substances and their metabolites are studied. The term has essentially the same meaning as pharmacokinetics, but the latter term should be restricted to the study of pharmaceutical substances.

Uncertainty factor (UF) A product of several single factors by which the NOAEL or LOAEL of the critical effect is divided to derive a tolerable intake. These factors account for adequacy of the pivotal study, interspecies extrapolation, inter-individual variability in humans, adequacy of the overall database, and nature of toxicity. The choice of UF should be based on the available scientific evidence.

CRITERIA FOR ENDPOINTS OTHER THAN CARCINOGENICITY

Criteria for selection of NOAEL/LOAEL

For those compounds reportedly without direct carcinogenic effects, determination of the highest concentration at which no adverse effects are observed, or the lowest concentration at which adverse effects are observed in humans, animals or plants is the first step in the derivation of the guideline value. This requires a thorough evaluation of available data on toxicity. The decision as to whether the LOAEL or the NOAEL should be used as a starting point for deriving a guideline value is mainly a matter of availability of data. If a series of data fixes both the LOAEL and the NOAEL, then either might be used. The gap between the lowest-observed-effect level and the no-observed-effect level is among the factors included in judgements concerning the appropriate uncertainty factor. Nevertheless, one needs to consider that in studies in experimental animals, the value of the NOAEL (or LOAEL) is an observed value that is dependent on the protocol and design of the study from which it was derived. There are several factors that influence the magnitude of the value observed, such as the species, sex, age, strain and developmental status of the animals studied; the group size; the sensitivity of the methods applied; and the selection of dose levels. Dose levels are frequently widely spaced, so that the observed NOAEL can be in some cases considerably less than the true no-adverse-effect level, and the observed LOAEL considerably higher than the true lowest-adverse-effect level *(1)*.

A single, free-standing no-observed-effect level that is not defined in reference to a lowest-observed-effect level or a LOAEL is not helpful. It is important to understand that, to be useful in setting guidelines, the NOAEL must be the highest level of exposure at which no adverse effects are detected. It is difficult to be sure that this has been identified unless the level of exposure at which adverse effects begin to appear has also been defined. Opinions on this subject differ, but the working consensus was that the level of exposure of concern in terms of human health is more easily related to the LOAEL, and this level was therefore used whenever possible. In the case of irritant and sensory effects on humans, it is desirable where possible to determine the no-observed-effect level. These effects are discussed in more detail below.

On the basis of the evidence concerning adverse effects, judgements about the uncertainty factors needed to minimize health risks were made. Averaging times were included in the specification of the guidelines, as the duration of exposure is often critical in determining toxicity. Criteria applied to each of these key factors are described below.

Criteria for selection of adverse effect

Definition of a distinction between adverse and non-adverse effects poses considerable difficulties. Any observable biological change might be considered an adverse effect under certain circumstances. An adverse effect has been defined as "any effect resulting in functional impairment and/or pathological lesions that may affect the performance of the whole organism or which contributes to a reduced ability to respond to an additional challenge" *(3)*. Even with such a definition, a significant degree of subjectivity and uncertainty remains. Ambient levels of major air pollutants frequently cause subtle effects that are typically detected only by sensitive methods. This makes it exceedingly difficult, if not impossible, to achieve a broad consensus as to which effects are adverse. To resolve this difficulty, it was agreed that the evidence should be ranked in three categories.

1. The first category comprises observations, even of potential health concern, that are single findings not verified by other groups. Because of the lack of verification by other investigators, such data could not readily be used as a basis for deriving a guideline value. They do, however, indicate the need for further research and may be considered in deriving an appropriate uncertainty factor based on the severity of the observed effects.

2. The second category is a lowest-observed-effect level (or no-observed-effect level) that is supported by other scientific information. When the results are in a direction that might result in pathological changes, there is a higher degree of health concern. Scientific judgement based on all available health information is used to determine how effects in this category can be used in determining the pollutant level that should be avoided so that excessive risk can be prevented.

3. The third category comprises levels of exposure at which there is clear evidence for substantial pathological changes; these findings have had a major influence on the derivation of the guidelines.

Benchmark approach

The benchmark dose (BMD) is the lower confidence limit of the dose that produces a given increase (e.g. 5% or 10%) in the level of an effect to which an uncertainty factor can be applied to develop a tolerable intake. It has a number of advantages over the NOAEL/LOAEL approach *(2)*. First, the BMD is derived on the basis of the entire dose–response curve for the critical, adverse effect rather than that from a single dose group as in the case of the NOAEL/LOAEL. Second, it can be calculated from data sets in which a NOAEL was not determined, eliminating the need for an additional uncertainty factor to be applied to the LOAEL. Third, definition of the BMD as a lower confidence limit accounts for the statistical power and quality of the data; that is, the confidence intervals around the dose–response curve for studies with small numbers of animals or of poor quality and thus lower statistical power would be wide, reflecting the greater uncertainty of the database. On the other hand, better studies would result in narrow confidence limits, and thus in higher BMDs.

Although there is no consensus on the incidence of effect to be used as basis for the BMD, it is generally agreed that the BMD should be comparable with a level of effect typically associated with the NOAEL or LOAEL. Allen et al. *(4, 5)* have estimated that a BMD calculated from the lower confidence limit at 5% is, on average, comparable to the NOAEL, whereas choosing a BMD at 10% is more representative of a LOAEL *(6)*. Choosing a BMD that is comparable to the NOAEL has two advantages: (*a*) it is within the experimental dose-range, eliminating the need to interpolate the dose–response curve to low levels; and (*b*) it justifies the application of similar uncertainty factors as are currently applied to the NOAEL for interspecies and intraspecies variation. It should be noted, however, that the main disadvantage of the benchmark approach is that it is not applicable for discrete toxicity data, such as histopathological or teratogenicity data.

Criteria for selection of uncertainty factors

In previous evaluations by WHO, uncertainty factors (sometimes called safety factors) have been applied to derive guidelines from evidence that conforms to accepted criteria for adverse effects on health *(7–9)*. Traditionally, the uncertainty (safety) factor has been used to allow for uncertainties in extrapolation from animals to humans and from a small group of individuals to a large population, including possibly undetected effects on particularly sensitive members of the population. In addition, uncertainty factors also account for possible synergistic effects of multiple exposures, the seriousness of the observed effects and the adequacy of existing data *(1)*. It is important to understand that the application of such factors does not indicate that it is known that humans are more sensitive than animal species but, rather, that the sensitivity of humans relative to that of other species is usually unknown. It is possible that humans are less sensitive than animals to some chemicals.

In this second edition of the air quality guidelines, the terms "safety factor" and "protection factor" have been replaced by the term "uncertainty factor". It is felt that this better explains the derivation and implications of such factors. Of course, such a factor is designed to provide an adequate level of protection and an adequate margin of safety, because these factors are applied in the derivation of guidelines for the protection of human health. They are not applied in the derivation of ecological guidelines because these already include a kind of uncertainty factor with regard to the variety of species covered.

A wide range of uncertainty factors are used in this second edition, based on scientific judgements concerning the interplay of various effects. The decision process for developing uncertainty factors has been complex, involving the transformation of mainly non-quantitative information into a single number expressing the judgement of a group of scientists.

Some of the factors taken into account in deciding the margin of protection can be grouped under the heading of scientific uncertainty. Uncertainty occurs because of limitations in the extent or quality of the database. One can confidently set a lower margin of protection (that is, use a smaller uncertainty factor) when a large number of high-quality, mutually supportive scientific experiments in different laboratories using different approaches clearly demonstrate the dose–response, including a lowest-observed-effect level and a no-observed-effect level. In reality, difficulties inherent in studying air pollutants, and the failure to perform much-needed and very specific research, means that often a large uncertainty factor has to be applied.

Where an uncertainty factor was adopted in the derivation of air quality guidelines, the reasoning behind the choice of this factor is given in the scientific background information. As previously mentioned, exceeding a guideline value with an incorporated uncertainty factor does not necessarily mean that adverse effects will result. Nevertheless, the risk to public health will increase, particularly in situations where the most sensitive population group is exposed to several pollutants simultaneously.

Individuals and groups within a population show marked differences in sensitivity to given pollutants. Individuals with pre-existing lung disease, for instance, may be at higher risk from exposure to air pollutants than healthy people. Differences in response can be due to factors other than pre-existing health, including age, sex, level of exercise taken and other unknown factors. Thus, the population must be considered heterogeneous in respect of response to air pollutants. This perhaps wide distribution of sensitivity combined with a distribution of exposure makes the establishment of population-based thresholds of effect very difficult. This problem is taken up in the section on particulate matter (page 186). Existing information tends not to allow adequate assessment of the proportion of the population that is likely to show an enhanced response. Nevertheless, an estimate of even a few percent of the total population entails a large number of people at increased risk.

Deriving a guideline from studies of effects on laboratory animals in the absence of human studies generally requires the application of an increased uncertainty factor, because humans may be more susceptible than laboratory animal species. Negative data from human studies will tend to reduce the magnitude of this uncertainty factor. Also of importance are the nature and reversibility of the reported effect. Deriving a guideline from data that show that a given level of exposure produces only slight alterations in physiological parameters requires a smaller uncertainty factor than when data showing a clearly adverse effect are used. Scientific judgement about uncertainty factors should also take into account the biochemical toxicology of pollutants, including the types of metabolite formed, the variability in metabolism or response in humans suggesting the existence of hypersusceptible groups, and the likelihood that the compound or its metabolites will accumulate in the body.

It is obvious, therefore, that diverse factors must be taken into account in proposing a margin of protection. The uncertainty factor cannot be assigned by a simple mathematical formula; it requires experience, wisdom and judgement.

Feasibility of adopting a standard approach

In preparing this second edition of the guidelines, the feasibility of developing a standard methodology for setting guidelines was discussed. It was agreed that Environmental Health Criteria No. 170 *(1)* was a valuable source of information. On the other hand, it was recognized that large variation in the data available for different compounds made the use of a standard approach impossible. Much of the difficulty concerns the adequacy of the database, and this has played a large part in controlling the methods of assessment adopted. This is illustrated in Table 1.

Table 1. Size and completeness of database in relation to assessment method

Examples	Completeness/ size of database	Uncertainties	Feasibility of expert judgement	Need for standardized approach
Nitrogen dioxide, ozone, lead	+++	+	+++	+
Manganese, nickel	++	++	++	++
Volatile organic compounds	+	+++	+	+++

It will be seen that when the database is strong (that is, when a good deal is known about the human toxicology of the compound) it is suggested that expert judgement can be used to set a guideline. In such circumstances the level of uncertainty is low. If, on the other hand, the database is weak, then a larger level of uncertainty will exist and there is much to be said for using a standardized approach, probably involving the application of a substantial uncertainty factor. The dangers of replacing expert judgement and the application of common sense with advanced, complex and sometimes not intuitively obvious statistical methods for deriving guidelines was discussed. It was agreed that a cautious approach should be adopted.

Criteria for selection of averaging times

The development of toxicity is a complex function of the interaction between concentration of a pollutant and duration of exposure. A chemical may cause acute, damaging effects after peak exposure for a short period and irreversible or incapacitating effects after prolonged exposure to lower concentrations. Our knowledge is usually insufficient to define accurately the relationship between effects on the one hand and concentration and time on the other. Expert judgement must be applied, therefore, based on the weight of the evidence available *(10)*.

Generally, when short-term exposures lead to adverse effects, short-term averaging times are recommended. The use of a long-term average under such conditions would be misleading, since the typical pattern of repeated peak exposures is lost during the averaging process and the risk manager would have difficulties in deciding on effective strategies. In other cases, knowledge of the exposure–response relationship may be sufficient to allow recommendation of a long averaging period. This is frequently the case for chemicals that accumulate in the body and thereby produce adverse effects. In such cases, the integral of concentration over a long period can have more impact than the pattern of peak exposure.

It should be noted that the specified averaging times are based on effects on health. Therefore, if the guidelines are used as a basis for regulation, the regulator needs to select the most appropriately and practically defined standards in relation to the guidelines, without necessarily adopting the guidelines directly. It was appreciated that monitoring techniques for some pollutants would not allow reporting of data in terms of the averaging times recommended in the guidelines. Under such circumstances, a compromise between the averaging time specified in the guidelines and that obtainable in practice has to be reached in setting an air quality standard.

A similar situation occurs for effects on vegetation. Plants are generally damaged by short-term exposures to high concentration as well as by long-term exposures to low concentration. Therefore, both short- and long-term guidelines to protect plants are proposed.

Criteria for consideration of sensory effects

Some of the substances selected for evaluation have malodorous properties at concentrations far below those at which toxic effects occur. Although odour annoyance cannot be regarded as an adverse health effect in a strict sense, it does affect the quality of life. Therefore, odour threshold levels have been indicated where relevant and used as a basis for separate guideline values.

For practical purposes, the following characteristics and respective levels were considered in the evaluation of sensory effects:

- intensity, where the *detection threshold level* is defined as the lower limit of the perceived intensity range (by convention the lowest concentration that can be detected in 50% of the cases in which it is present);

- quality, where the *recognition threshold level* is defined as the lowest concentration at which the sensory effect, such as odour, can be recognized correctly in 50% of the cases; and

- acceptability and annoyance, where the *nuisance threshold level* is defined as the concentration at which not more than a small proportion of the population (less than 5%) experiences annoyance for a small part of the time (less than 2%); since annoyance will be influenced by a number of psychological and socioeconomic factors, a nuisance threshold level cannot be defined on the basis of concentration alone.

During revision of the guidelines, the problems of irritation (for example, of the skin) and headache were also considered as possible problems of annoyance. It was agreed that headache should be regarded as a health endpoint and not merely as a matter of annoyance.

CRITERIA FOR CARCINOGENIC ENDPOINT

Cancer risk assessment is basically a two-step procedure, involving a qualitative assessment of how likely it is that an agent is a human carcinogen, and a quantitative assessment of the cancer risk that is likely to occur at given levels and duration of exposure (*11*).

Qualitative assessment of carcinogenicity

The decision to consider a substance as a carcinogen is based on the qualitative evaluation of all available information on carcinogenicity, ensuring that the association is unlikely to be due to chance alone. Here the classification criteria of the International Agency for Research on Cancer (IARC) have been applied (Box 1). In dealing with carcinogens, a "general rule" and exceptions from this were defined. The "general rule" states that for compounds in IARC Groups 1 and 2A (proven human carcinogens, and carcinogens with at least limited evidence of human carcinogenicity), guideline values are derived with the use of quantitative risk assessment with low-dose risk extrapolation. For compounds in Groups 2B (inadequate evidence in humans but sufficient evidence in animals), 3 (unclassifiable as to carcinogenicity in humans) and 4 (noncarcinogenic), guideline values are derived with the use of a threshold (uncertainty factor) method. For compounds in Group 2B, this may incorporate a separate factor for the possibility of a carcinogenic effect in humans.

In case of sufficient scientific evidence, one may be justified in deviating from the "general rule". First, a compound classified in Group 1 or 2A may be assessed with the use of the uncertainty factor methodology, provided that there is strong evidence that it is not genotoxic as judged from a battery of short-term test systems for gene mutation, DNA damage, etc. In such cases it can be established with certainty that an increase in exposure to the compound is associated with an increase in cancer incidence only above a certain level of exposure. It was considered that this required a level of understanding of the mechanisms of action not presently available for the compounds classified as Group 1 or 2A on the current list. Second, a compound in Group 2B may be assessed with the use of quantitative risk assessment methods instead of the uncertainty factor approach. This may be considered appropriate where the mechanism of carcinogenesis in animals is likely to be a non-threshold phenomenon as indicated, for example, by the genotoxic activity of the compound in different short-term test systems.

Quantitative assessment of carcinogenic potency

The aim of quantitative risk assessment is to use information available from very specific study situations to predict the risk to the general population posed by exposure to ambient levels of carcinogens. In general, therefore, quantitative risk assessment includes the extrapolation of risk from relatively high dose levels (characteristic of animal experiments or occupational exposures), where cancer responses can be measured, to relatively low dose levels, which are of concern in environmental protection and where such

Box 1. Classification criteria of the International Agency for Research on Cancer

Group 1 – the agent (mixture) is carcinogenic to humans.
The exposure circumstance entails exposures that are carcinogenic to humans.
This category is used when there is *sufficient evidence* of carcinogenicity in humans. Exceptionally, an agent (mixture) may be placed in this category when evidence in humans is less than sufficient but there is *sufficient evidence* of carcinogenicity in experimental animals and strong evidence in exposed humans that the agent (mixture) acts through a relevant mechanism of carcinogenicity.

Group 2
This category includes agents, mixtures and exposure circumstances for which, at one extreme, the degree of evidence of carcinogenicity in humans is almost sufficient, as well as those for which, at the other extreme, there are no human data but for which there is evidence of carcinogenicity in experimental animals. Agents, mixtures and exposure circumstances are assigned to either group 2A (probably carcinogenic to humans) or group 2B (possibly carcinogenic to humans) on the basis of epidemiological and experimental evidence of carcinogenicity and other relevant data.

Group 2A – the agent (mixture) is probably carcinogenic to humans.
The exposure circumstance entails exposures that are probably carcinogenic to humans.
This category is used when there is *limited evidence* of carcinogenicity in humans and sufficient evidence of carcinogenicity in experimental animals. In some cases, an agent (mixture) may be classified in this category when there is inadequate evidence of carcinogenicity in humans and *sufficient evidence* of carcinogenicity in experimental animals and strong evidence that the carcinogenesis is mediated by a mechanism that also operates in humans. Exceptionally, an agent, mixture or exposure circumstance may be classified in this category solely on the basis of limited evidence of carcinogenicity in humans.

Group 2B – the agent (mixture) is possibly carcinogenic to humans.
The exposure circumstance entails exposures that are possibly carcinogenic to humans.
This category is used for agents, mixtures and exposure circumstances for which there is *limited evidence* of carcinogenicity in humans and less than *sufficient evidence of* carcinogenicity in experimental animals. It may also be used when there is *inadequate evidence* of carcinogenicity in humans but there is *sufficient evidence* of carcinogenicity in experimental animals. In some instances, an agent, mixture or exposure circumstance for which there is *inadequate evidence* of carcinogenicity in humans but *limited evidence* of carcinogenicity in experimental animals together with supporting evidence from other relevant data may be placed in this group.

Box 1. (contd)

Group 3 – The agent (mixture or exposure circumstance) is not classifiable as to its carcinogenicity to humans.
This category is used most commonly for agents, mixtures and exposure circumstances for which the evidence of carcinogenicity is inadequate in humans and inadequate or limited in experimental animals. Exceptionally, agents (mixtures) for which the evidence of carcinogenicity is inadequate in humans but sufficient in experimental animals may be placed in this category when there is strong evidence that the mechanism of carcinogenicity in experimental animals does not operate in humans. Agents, mixtures and exposure circumstances that do not fall into any other group are also placed in this category.

Group 4 – The agent (mixture) is probably not carcinogenic to humans.
This category is used for agents or mixtures for which there is *evidence suggesting lack of carcinogenicity* in humans and in experimental animals. In some instances, agents or mixtures for which there is *inadequate evidence* of carcinogenicity in humans but *evidence suggesting lack of carcinogenicity* in experimental animals, consistently and strongly supported by a broad range of other relevant data, may be classified in this group.

Source: IARC *(12)*.

risks are too small to be measured directly, either by animal studies or by epidemiological studies.

The choice of the extrapolation model depends on the current understanding of the mechanisms of carcinogenesis *(13)*, and *no* single mathematical procedure can be regarded as fully appropriate for low-dose extrapolation. Methods based on a linear, non-threshold assumption have been used at the national and international level more frequently than models that assume a safe or virtually safe threshold.

In these guidelines, the risk associated with lifetime exposure to a certain concentration of a carcinogen in the air has been estimated by linear extrapolation and the carcinogenic potency expressed as the incremental unit risk estimate. The incremental unit risk estimate for an air pollutant is defined as "the additional lifetime cancer risk occurring in a hypothetical population in which all individuals are exposed continuously from birth throughout their lifetimes to a concentration of 1 $\mu g/m^3$ of the agent in the air they breathe" *(14)*.

The results of calculations expressed in unit risk estimates provide the opportunity to compare the carcinogenic potency of different compounds and can help to set priorities in pollution control, taking into account current levels of exposure. By using unit risk estimates, any reference to the "acceptability" of risk is avoided. The decision on the acceptability of a risk should be made by national authorities within the framework of risk management. To support authorities in the decision-making process, the guideline sections for carcinogenic pollutants provide the concentrations in air associated with an excess cancer risk of 1 in a population of 10 000, 1 in 100 000 or 1 in 1 000 000, respectively, calculated from the unit risk.

For those substances for which appropriate human studies are available, the method known as the "average relative risk model" has been used, and is therefore described in more detail below.

Several methods have been used to estimate the incremental risks based on data from animal studies. Two general approaches have been proposed. A strictly linearized estimate has generally been used by the US Environmental Protection Agency (EPA) *(14)*. Nonlinear relations have been proposed for use when the data derived from animal studies indicate a nonlinear dose–response relationship or when there is evidence that the capacity to metabolize the polluting chemical to a carcinogenic form is of limited capacity.

Quantitative assessment of carcinogenicity based on human data

Quantitative assessment using the average relative risk model comprises four steps: (*a*) selection of studies; (*b*) standardized description of study results in terms of relative risk, exposure level and duration of exposure; (*c*) extrapolation towards zero dose; and (*d*) application to a general (hypothetical) population.

First, a reliable human study must be identified, where the exposure of the study population can be estimated with acceptable confidence and the excess cancer incidence is statistically significant. If several studies exist, the best representative study should be selected or several risk estimates evaluated.

Once a study is identified, the relative risk as a measure of response is calculated. It is important to note that the 95% confidence limits around the central estimate of the relative risk can be wide and should be specifically stated and evaluated. The relative risk is then used to calculate the excess lifetime cancer risk expressed as unit risk (UR) associated with a lifetime exposure to 1 $\mu g/m^3$, as follows:

$$UR = \frac{P_0(RR - 1)}{X}$$

where: P_0 = background lifetime risk; this is taken from age/cause-specific death or incidence rates found in national vital statistics tables using the life table methodology, or it is available from a matched control population

RR = relative risk, being the ratio between the observed (O) and expected (E) number of cancer cases in the exposed population; the relative risk is sometimes expressed as the standardized mortality ratio SMR = (O/E) × 100

X = lifetime average exposure (standardized lifetime exposure for the study population on a lifetime continuous exposure basis); in the case of occupational studies, X represents a conversion from the occupational 8-hour, 240-day exposure over a specific number of working years and can be calculated as X = 8-hour TWA × 8/24 × 240/365 × (average exposure duration [in years])/(life expectancy [70 years]), where TWA is the time-weighted average (μg/m³).

It should be noted that the unit lifetime risk depends on P_0 (background lifetime risk), which is determined from national age-specific cancer incidence or mortality rates. Since these rates are also determined by exposures other than the one of interest and may vary from country to country, it follows that the UR may also vary from one country to another.

Necessary assumptions for average relative risk method

Before any attempt is made to assess the risk in the general population, numerous assumptions are needed at each phase of the risk assessment process to fill in various gaps in the underlying scientific database. As a first step in any given risk assessment, therefore, an attempt should be made to identify the major assumptions that have to be made, indicating their probable consequences. These assumptions are as follows.

1. *The response (measured as relative risk) is some function of cumulative dose or exposure.*

2. *There is no threshold dose for carcinogens.*
Many stages in the basic mechanism of carcinogenesis are not yet known or are only partly understood. Taking available scientific findings into consideration,

however, several scientific bodies *(8, 15–17)* have concluded that there is no scientific basis for assuming a threshold or no-effect level for chemical carcinogens. This view is based on the fact that most agents that cause cancer also cause irreversible damage to deoxyribonucleic acid (DNA). The assumption applies for all non-threshold models.

3. *The linear extrapolation of the dose–response curve towards zero gives an upper-bound conservative estimate of the true risk function if the unknown (true) dose–response curve has a sigmoidal shape.*
The scientific justification for the use of a linear non-threshold extrapolation model stems from several sources: the similarity between carcinogenesis and mutagenesis as processes that both have DNA as target molecules; the strong evidence of the linearity of dose–response relationships for mutagenesis; the evidence for the linearity of the DNA binding of chemical carcinogens in the liver and skin; the evidence for the linearity in the dose–response relationship in the initiation stage of the mouse 2-stage tumorigenesis model; and the rough consistency with the linearity of the dose–response relationships for several epidemiological studies. This assumption applies for all linear models.

4. *There is constancy of the relative risk in the specific study situation.*
In a strict sense, constancy of the relative risk means that the background age/cause-specific rate at any time is increased by a constant factor. The advantage of the average relative risk method is that this needs to be true only for the average.

Advantages of the method

The average relative risk method was selected in preference to many other more sophisticated extrapolation models because it has several advantages, the main one being that it seems to be appropriate for a fairly large class of different carcinogens, as well as for different human studies. This is possible because averaging doses, that is, averaging done over concentration and duration of exposure, gives a reasonable measure of exposure when dose rates are not constant in time. This may be illustrated by the fact that the use of more sophisticated models *(14, 18, 19)* results in risk estimates very similar to those obtained by the average relative risk method.

Another advantage of the method is that the carcinogenic potency can be calculated when estimates of the average level and duration of exposure are the only known parameters besides the relative risk. Furthermore, the method has the advantage of being simple to apply, allowing non-experts in the field of risk models to calculate a lifetime risk from exposure to the carcinogens.

Limitations of the method

As pointed out earlier, the average relative risk method is based on several assumptions that appear to be valid in a wide variety of situations. There are specific situations, however, in which the method cannot be recommended, mainly because the assumptions do not hold true.

The cumulative dose concept, for instance, is inappropriate when the mechanism of the carcinogen suggests that it cannot produce cancer throughout all stages of the cancer development process. Also, specific toxicokinetic properties, such as a higher excretion rate of a carcinogen at higher doses or a relatively lower production rate of carcinogenic metabolites at lower doses, may diminish the usefulness of the method in estimating cancer risk. Furthermore, supralinearity of the dose–response curve or irregular variations in the relative risk over time that cannot be eliminated would reduce the value of the model. Nevertheless, evidence concerning these limitations either does not exist or is still too preliminary to make the average relative risk method inappropriate for carcinogens evaluated here.

A factor of uncertainty, rather than of methodological limitation, is that data on past exposure are nearly always incomplete. Although it is generally assumed that in the majority of studies the historical dose rate can be determined within an order of magnitude, there are possibly greater uncertainties, even of more than two orders of magnitude, in some studies. In the risk assessment process it is of crucial importance that this degree of uncertainty be clearly stated. This is often done simply by citing upper and lower limits of risk estimates. Duration of exposure and the age- and time-dependence of cancer caused by a particular substance are less uncertain parameters, although the mechanisms of relationship are not so well understood *(11)*.

Risk estimates from animal cancer bioassays

Animal bioassays of chemicals provide important information on the human risk of cancer from exposure to chemicals. These data enhance our confidence in assessing human cancer risks on the basis of epidemiological data.

There is little doubt of the importance of animal bioassay data in reaching an informed decision on the carcinogenic potential of a chemical. The collection and use of data such as those on saturation mechanisms, absorption, distribution and metabolic pathways, as well as on interaction with other chemicals, is important and should be continued. Regrettably, these data were not always available for the air pollutants evaluated during the update and revision of the guidelines. The process of evaluating guidelines and the impact of exposure to these chemicals on human health should continue and be revised as new information becomes available.

Several chemicals considered in this publication have been studied using animal cancer bioassays. The process is continuing and new information on the potential carcinogenicity of chemicals is rapidly appearing. Consequently, the status of chemicals is constantly being reassessed.

There is no clear consensus on appropriate methodology for the risk assessment of chemicals for which the critical effect may not have a threshold, such as genotoxic carcinogens and germ cell mutagens. A number of approaches based largely on characterization of dose–response have been adopted for assessment of such effects:

- quantitative extrapolation by mathematical modelling of the dose–response curve to estimate the risk at likely human intakes or exposures (low-dose risk extrapolation);
- relative ranking of potencies in the experimental range; and
- division of effect levels by an uncertainty factor.

Low-dose risk extrapolation has been accomplished by the use of mathematical models such as the Armitage-Doll multi-stage model. In more recently developed biological models, the different stages in the process of carcinogenesis have been incorporated and time to tumour has been taken into account *(20)*. In some cases, such as that of butadiene, uncertainty regarding the metabolism in humans and experimental animals precluded the choice of the appropriate animal model for low-dose risk extrapolation. In other cases where data permitted, attempts were made to incorporate the dose delivered to the target tissue into the dose–response analysis (physiologically based pharmacokinetic modelling).

During revision of the guidelines, other approaches to establishing guideline levels for carcinogens were considered. Such approaches involve the identification of a level of exposure at which the risk is known to be small and the application of uncertainty factors to derive a level of exposure at which the risk is accepted as being exceedingly small or negligible. This approach has been adopted in the United Kingdom, for example. It was agreed that such an approach might be applicable on a national or smaller scale, but that it was unlikely to be generally applicable.

Interpretation of risk estimates

The risk estimates presented in this book should *not* be regarded as being equivalent to the true cancer risk. It should be noted that crude expression of risk in terms of excess incidence or numbers of cancers per unit of the population at doses or concentrations much less than those on which the estimates are based may be inappropriate, owing to the uncertainties of quantitative extrapolation over several orders of magnitude. Estimated risks are believed to represent only the plausible upper bounds, and may vary widely depending on the assumptions on which they are based.

The presented quantitative risk estimates can provide policy-makers with rough estimates of risk that may serve well as a basis for setting priorities, balancing risks and benefits, and establishing the degree of urgency of public health problems among subpopulations inadvertently exposed to carcinogens. A risk management approach for compounds for which the critical effect is considered not to have a threshold involves eliminating or reducing exposure as far as practically or technologically possible. Characterization of the dose–response, as indicated in the procedures described above, can be used in conjunction with this approach to assess the need to reduce exposure.

Combined exposures

Exposure to combinations of air pollutants is inevitable. Data dealing with the effects of co-exposure to air pollutants are, however, very limited and it is not possible to recommend guidelines for such combinations. Of course, measures taken to control air pollution frequently lead to the reduction in concentrations of more than one pollutant. This is often achieved by controlling sources of pollutants rather than by focusing on individual pollutants.

ECOLOGICAL EFFECTS

The importance of taking an integrated view of both health and ecological effects in air quality management was recognized from the beginning of the project. Ecological effects may have a significant indirect influence on human health and wellbeing. For example, most of the major urban air pollutants are known to have adverse effects at low levels on plants, including food crops. A consultation group was therefore convened to consider the ecological effects on terrestrial vegetation of sulfur dioxide, nitrogen dioxide, and ozone and other photochemical oxidants. These substances are important both because of the high anthropogenic amounts produced and because of their wide distribution. They deserve special attention because of significant adverse effects on ecological systems in concentrations far below those known to be harmful to humans.

The pollutants selected for consideration here form only part of the vast range of air pollutants that have ecological effects. The project timetable permitted only an evaluation of adverse effects on terrestrial plant life, although effects on animal and aquatic ecosystems are also of great concern in parts of Europe. Nevertheless, even this limited evaluation clearly indicates the importance attached to the ecological effects of such pollutants in the European Region.

REFERENCES

1. *Assessing human health risks of chemicals: derivation of guidance values for health-based exposure limits.* Geneva, World Health Organization, 1994 (Environmental Health Criteria, No. 170).
2. CRUMP, K.S. A new method for determining allowable daily intakes. *Fundamental and applied toxicology*, **4**: 854–871 (1984).
3. US ENVIRONMENTAL PROTECTION AGENCY. Guidelines and methodology used in the preparation of health effect assessment chapters of the consent decree water quality criteria. *Federal register,* **45**: 79347–79357 (1980).
4. ALLEN, B.C. ET AL. Dose–response modeling for developmental toxicity. *Toxicologist,* **12**: 300 (1992).
5. ALLEN, B.C. ET AL. Comparison of quantitative dose response modeling approaches for evaluating fetal weight changes in segment II developmental toxicity studies. *Teratology,* **47**(5): 41 (1993).
6. FARLAND, W. & DOURSON, M. Noncancer health endpoints: approaches to quantitative risk assessment. *In*: Cothern, R., ed. *Comparative environmental risk assessment.* Boca Raton, FL, Lewis Publishers, 1992, pp. 87–106.
7. VETTORAZZI, G. *Handbook of international food regulatory toxicology. Vol. 1. Evaluations.* New York, SP Medical and Scientific Books, 1980.
8. *Guidelines for drinking-water quality. Vol. 1. Recommendations.* Geneva, World Health Organization, 1984.
9. *Air quality guidelines for Europe*. Copenhagen, WHO Regional Office for Europe, 1987 (WHO Regional Publications, European Series, No. 23).
10. *Air quality criteria and guides for urban air pollutants: report of a WHO Expert Committee*. Geneva, World Health Organization, 1972 (WHO Technical Report Series, No. 506).
11. PEAKALL, D.B. ET AL. Methods for quantitative estimation of risk from exposure to chemicals. *In*: Vouk, V.B. et al., ed. *Methods for estimating risk of chemical injury: human and non-human biota and ecosystems.* New York, John Wiley & Sons, 1985.
12. *Polychlorinated dibenzo*-para-*dioxins and polychlorinated dibenzofurans.* Lyons, International Agency for Research on Cancer, 1997 (IARC Monographs on the Evaluation of Carcinogenic Risks to Humans, Vol. 69).

13. ANDERSON, E.L. Quantitative approaches in use in the United States to assess cancer risk. *In*: Vouk, V.B. et al., ed. *Methods for estimating risk of chemical injury: human and non-human biota and ecosystems.* New York, John Wiley & Sons, 1985.
14. *Health assessment document for nickel.* Research Triangle Park, NC, US Environmental Protection Agency, 1985 (Final Report No. EPA-600/8-83-12F).
15. ANDERSON, E.L. ET AL. Quantitative approaches in use to assess cancer risk. *Risk analysis,* **3**: 277–295 (1983).
16. NATIONAL RESEARCH COUNCIL. *Drinking water and health.* Washington, DC, National Academy Press, 1977.
17. *Risk assessment and risk management of toxic substances. A report to the Secretary, Department of Health and Human Services.* Washington, DC, US Department of Health and Human Services, 1985.
18. *Health assessment document for chromium.* Washington, DC, US Environmental Protection Agency, 1984 (Final report EPA-600-8-83-014F).
19. *Health assessment document for inorganic arsenic.* Washington, DC, US Environmental Protection Agency, 1984 (Final report EPA-600-8-83-021F).
20. MOOLGAVKAR, S.H. ET AL. A stochastic two-stage model for cancer risk assessment. I. The hazard function and the probability of tumor. *Risk analysis,* **8**: 383–392 (1988).

CHAPTER 3

Summary of the guidelines

The term "guidelines" in the context of this book implies not only numerical values (guideline values), but also any kind of guidance given. Accordingly, for some substances the guidelines encompass recommendations of a more general nature that will help to reduce human exposure to harmful levels of air pollutants. For some pollutants no guideline values are recommended, but risk estimates are indicated instead.

The numerical guideline values and the risk estimates for carcinogens (Tables 2–4) should be regarded as the shortest possible summary of a complex scientific evaluation process. Nevertheless, the information given in the tables should not be used without reference to the rationale given in the chapters on the respective pollutants. Scientific results are an abstraction of real situations, and this is even more true for numerical values and risk estimates based on such results. Numerical guideline values, therefore, are not to be regarded as separating the acceptable from the unacceptable, but rather as indications. They are proposed in order to help avoid major discrepancies in reaching the goal of effective protection against recognized hazards for human health and the environment. Moreover, numerical

Table 2. Guideline values for individual substances based on effects other than cancer or odour/annoyance

Substance	Time-weighted average	Averaging time
Cadmium	5 ng/m^3 [a]	annual
Carbon disulfide[b]	100 µg/m^3	24 hours
Carbon monoxide	100 mg/m^3 [c]	15 minutes
	60 mg/m^3 [c]	30 minutes
	30 mg/m^3 [c]	1 hour
	10 mg/m^3	8 hours
1,2-Dichloroethane[b]	0.7 mg/m^3	24 hours
Dichloromethane	3 mg/m^3	24 hours
	0.45 mg/m^3	1 week

Table 2 (contd)

Substance	Time-weighted average	Averaging time
Fluoride[d]	–	–
Formaldehyde	0.1 mg/m^3	30 minutes
Hydrogen sulfide[b]	150 µg/m^3	24 hours
Lead	0.5 µg/m^3	annual
Manganese	0.15 µg/m^3	annual
Mercury	1 µg/m^3	annual
Nitrogen dioxide	200 µg/m^3	1 hour
	40 µg/m^3	annual
Ozone	120 µg/m^3	8 hours
Particulate matter[e]	Dose–response	–
Platinum[f]	–	–
PCBs[g]	–	–
PCDDs/PCDFs[h]	–	–
Styrene	0.26 mg/m^3	1 week
Sulfur dioxide	500 µg/m^3	10 minutes
	125 µg/m^3	24 hours
	50 µg/m^3	annual
Tetrachloroethylene	0.25 mg/m^3	annual
Toluene	0.26 mg/m^3	1 week
Vanadium[b]	1 µg/m^3	24 hours

[a] The guideline value is based on the prevention of a further increase of cadmium in agricultural soils, which is likely to increase the dietary intake.

[b] Not re-evaluated for the second edition of the guidelines.

[c] Exposure at these concentrations should be for no longer than the indicated times and should not be repeated within 8 hours.

[d] Because there is no evidence that atmospheric deposition of fluorides results in significant exposure through other routes than air, it was recognized that levels below 1 µg/m^3, which is needed to protect plants and livestock, will also sufficiently protect human health.

[e] The available information for short- and long-term exposure to PM_{10} and $PM_{2.5}$ does not allow a judgement to be made regarding concentrations below which no effects would be expected. For this reason no guideline values have been recommended, but instead risk estimates have been provided (see Chapter 7, Part 3).

[f] It is unlikely that the general population, exposed to platinum concentrations in ambient air at least three orders of magnitude below occupational levels where effects were seen, may develop similar effects. No specific guideline value has therefore been recommended.

[g] No guideline value has been recommended for PCBs because inhalation constitutes only a small proportion (about 1–2%) of the daily intake from food.

[h] No guideline value has been recommended for PCDDs/PCDFs because inhalation constitutes only a small proportion (generally less than 5%) of the daily intake from food.

guidelines for different substances are not directly comparable. Variations in the quality and extent of the scientific information and in the nature of critical effects, although usually reflected in the applied uncertainty factor, result in guideline values that are only to a limited extent comparable between pollutants.

Owing to the different bases for evaluation, the numerical values for the various air pollutants should be considered in the context of the accompanying scientific documentation giving the derivation and scientific considerations. Any *isolated* interpretation of numerical data should therefore be avoided, and guideline values should be used and interpreted in conjunction with the information contained in the appropriate sections.

It is important to note that the approach taken in the preparation of the guidelines was to evaluate data on the health effects of individual compounds. Consequently, each chemical was considered in isolation. Pollutant mixtures can yield different toxic effects, but data are at present insufficient for guidelines relating to mixtures to be laid down. There is little emphasis on interaction between pollutants that might lead to additive or synergistic effects and on the environmental fate of pollutants, though there is growing evidence about the role of solvents in atmospheric photochemical processes leading to the formation or degradation of ozone, the formation of acid rain, and the propensity of metals and trace elements to accumulate in environmental niches. These factors militate strongly against allowing a rise in ambient pollutant levels. Many uncertainties still remain, particularly regarding the ecological effects of pollutants, and therefore efforts should be continued to maintain air quality at the best possible level.

GUIDELINE VALUES BASED ON NON-CANCER EFFECTS OTHER THAN CANCER

The guideline values for individual substances based on effects other than cancer and annoyance from odour are given in Table 2. The emphasis in the guidelines is placed on exposure, since this is the element that can be controlled to lessen dose and hence lessen the consequent health effect. When general ambient air levels are orders of magnitude lower than the guideline values, present exposures are unlikely to cause concern. Guideline values in those cases are directed only to specific release episodes or specific indoor pollution problems.

As stated earlier, the starting point for the derivation of guideline values was to define the lowest concentration at which adverse effects are observed. On the basis of the body of scientific evidence and judgements of uncertainty factors, numerical guideline values were established to the extent possible. Compliance with the guideline values does not, however, guarantee the absolute exclusion of undesired effects at levels below the guideline values. It means only that guideline values have been established in the light of current knowledge and that uncertainty factors based on the best scientific judgements have been incorporated, though some uncertainty cannot be avoided.

For some of the substances, a direct relationship between concentrations in air and possible toxic effects is very difficult to establish. This is especially true of those pollutants for which a greater body burden results from ingestion than from inhalation. For instance, available data show that for the general population the food chain is the critical route of non-occupational exposure to lead and cadmium, and to persistent organic pollutants such as dioxins and PCBs. On the other hand, emissions of these pollutants into air may contribute significantly to the contamination of food by these compounds. Complications of this kind were taken into consideration, and an attempt was made to develop guidelines that would also prevent those toxic effects of air pollutants that resulted from uptake by both ingestion and inhalation.

For certain compounds, such as organic solvents, the proposed health-related guidelines are orders of magnitude higher than current ambient levels. The fact that existing environmental levels for some substances are much lower than the guideline levels by no means implies that pollutant burdens may be increased up to the guideline values. Any level of air pollution is a matter of concern, and the existence of guideline values never means a licence to pollute.

Unfortunately, the situation with regard to actual environmental levels and proposed guideline values for some substances is just the opposite – guideline values are below existing levels in some parts of Europe. For instance, the guideline values recommended for major urban air pollutants such as nitrogen dioxide, ozone and sulfur dioxide point to the need for a significant reduction of emissions in some areas.

For substances with malodorous properties at concentrations below those where toxic effects occur, guideline values likely to protect the public from odour nuisance were established; these were based on data provided by expert panels and field studies (Table 3). In contrast to other air pollutants, odorous substances in ambient air often cannot be determined easily and

Table 3. Rationale and guideline values based on sensory effects or annoyance reactions, using an averaging time of 30 minutes

Substance	Detection threshold	Recognition threshold	Guideline value
Carbon disulfide[a] (index substance for viscose emissions)	200 μg/m³	–	20 μg/m³
Hydrogen sulfide[a]	0.2–2.0 μg/m³	0.6–6.0 μg/m³	7 μg/m³
Formaldehyde	0.03–0.6 mg/m³	–	0.1 mg/m³
Styrene	70 μg/m³	210–280 μg/m³	70 μg/m³
Tetrachloroethylene	8 mg/m³	24–32 mg/m³	8 mg/m³
Toluene	1 mg/m³	10 mg/m³	1 mg/m³

[a] Not re-evaluated for the second edition of the guidelines.

systematically by analytical methods because the concentrations are usually very low. Furthermore, odours in the ambient air frequently result from a complex mixture of substances and it is difficult to identify individual ones; future work may have to concentrate on odours as perceived by individuals rather than on separate odorous substances.

GUIDELINES BASED ON CARCINOGENIC EFFECTS

In establishing criteria upon which guidelines could be based, it became apparent that carcinogens and noncarcinogens would require different approaches. These approaches are determined by theories of carcinogenesis, which postulate that there is no threshold for effects (that is, that there is no safe level). Risk managers are therefore faced with two choices: either to prohibit a chemical or to regulate it at levels that result in an acceptable degree of risk. Indicative figures for risk and exposure assist the risk manager to reach the latter decision. Air quality guidelines are therefore indicated in terms of incremental unit risks (Table 4) in respect of those carcinogens that are considered to be genotoxic (see Chapter 2). To allow risk managers to judge the acceptability of risks, this edition of the guidelines has provided concentrations of carcinogenic air pollutants associated with an excess lifetime cancer risk of 1 per 10 000, 1 per 100 000 and 1 per 1 000 000.

For butadiene, there is substantial information on its mutagenic and carcinogenic activity. It has been shown that butadiene is mutagenic in both

Table 4. Carcinogenic risk estimates based on human studies[a]

Substance	IARC Group	Unit risk[b]	Site of tumour
Acrylonitrile[c]	2A	2×10^{-5}	lung
Arsenic	1	1.5×10^{-3}	lung
Benzene	1	6×10^{-6}	blood (leukaemia)
Butadiene	2A	—	multisite
Chromium (VI)	1	4×10^{-2}	lung
Nickel compounds	1	4×10^{-4}	lung
Polycyclic aromatic hydrocarbons (BaP)[d]	—	9×10^{-2}	lung
Refractory ceramic fibres	2B	1×10^{-6} (fibre/l)$^{-1}$	lung
Trichloroethylene	2A	4.3×10^{-7}	lung, testis
Vinyl chloride[c]	1	1×10^{-6}	liver and other sites

[a] Calculated with average relative risk model.

[b] Cancer risk estimates for lifetime exposure to a concentration of 1 μg/m³.

[c] Not re-evaluated for the second edition of the guidelines.

[d] Expressed as benzo[*a*]pyrene (based on a benzo[*a*]pyrene concentration of 1 μg/m³ in air as a component of benzene-soluble coke-oven emissions).

bacterial and mammalian systems, but metabolic activation into DNA-reactive metabolites is required for this activity. In general, metabolism of butadiene to epoxides in humans is significantly less than in mice and rats, with mice having the highest metabolic activity. Human cancer risk estimates for butadiene based on bioassays vary considerably depending on the animal species used, with risk estimates based on data in mice being 2–3 orders of magnitude higher than those based on rat data. At present, no definite conclusion can be made as to which animal species is most appropriate for human cancer risk estimates, and thus no guideline value is recommended for butadiene.

Separate consideration is given to risk estimates for asbestos (Table 5) and radon daughters (Table 6) because they refer to different physical units, and the risk estimates are indicated in the form of ranges.

Risk estimation for residential radon exposure has often been based on extrapolation of findings in underground miners. Several circumstances, however, make such estimates uncertain for the general population: exposure to other factors in the mines; differences in age and sex; size distribution of aerosols; the attached fraction of radon progeny; breathing rate; and

Table 5. Risk estimates for asbestos

Concentration	Range of lifetime risk estimates	
500 F*/m^3 (0.0005 F/ml)[a]	10^{-6}–10^{-5}	(lung cancer in a population where 30% are smokers)
	10^{-5}–10^{-4}	(mesothelioma)

[a] F* = fibres measured by optical methods.

Table 6. Risk estimates and recommended action level for radon progeny

Exposure	Lung cancer excess lifetime risk estimate	Recommended level for remedial action in buildings
1 Bq/m^3	3–6 × 10^{-5}	≥ 100 Bq/m^3 (annual average)

route. Furthermore, uncertainties in the exposure–response exist, and possible differences in the relative risk estimates for smokers and non-smokers are not fully understood (see Chapter 8, Part 3).

For radon, a unit risk of approximately 3–6 × 10^{-5} per Bq/m^3 can be calculated assuming a life time risk of lung cancer of 3% (Table 6). This means that a person living in an average European house with 50 Bq/m^3 has a lifetime excess lung cancer risk of 1.5–3 × 10^{-3} Thus current levels of radon in dwellings and other buildings are of public health concern. In addition it should be noted that a lifetime lung cancer risk below about 10^{-4} could normally not be expected to be achievable because natural concentration of radon in ambient air outdoors is about 10 Bq/m^3. Therefore no numerical guideline value for radon is recommended.

It is important to note that quantitative risk estimates may give an impression of accuracy that they do not in fact have. An excess of cancer in a population is a biological effect and not a mathematical function, and uncertainties of risk estimation are caused not only by inadequate exposure data but also, for instance, by the fact that specific metabolic properties of agents are not reflected in the models. The guidelines do not indicate therefore that a specified lifetime risk is virtually safe or acceptable.

The decision on the acceptability of a certain risk should be taken by the national authorities in the context of a broader risk management process. Risk estimate figures should not be applied in isolation when regulatory decisions are being made; combined with data on exposure levels and individuals exposed, they may be a useful contribution to risk assessment. Risk assessment can then be used together with technological, economic and other considerations in the risk management process.

GUIDELINES BASED ON EFFECTS ON VEGETATION

Although the main objective of the air quality guidelines is the direct protection of human health, it was decided that ecological effects of air pollutants on vegetation should also be considered. The effects of air pollutants on the natural environment are of special concern when they occur at concentrations lower than those that damage human health. In such cases, air quality guidelines based only on effects on human health would not allow for environmental damage that might indirectly affect human wellbeing.

Ecologically based guidelines for preventing adverse effects on terrestrial vegetation were included in the first edition of this book, and guidelines were recommended for some gaseous air pollutants. Since that time, however, significant advances in the scientific understanding of the impacts of air pollutants on the environment have been made. For the updating and revision of the guidelines, the ecological effects of major air pollutants were considered in more detail within the framework of the Convention on Long-range Transboundary Air Pollution. This capitalizes on the scientific work undertaken since 1988 to formulate criteria for the assessment of the effects of air pollutants on the natural environment, such as critical levels and critical loads.

It should be understood that the pollutants selected (SO_2, NO_x and ozone/photochemical oxidants) (Table 7) are only a few of a larger category of air

Table 7. Guideline values for individual substances based on effects on terrestrial vegetation

Substance	Guideline value	Averaging time
SO_2: critical level	10–30 μg/m^3 [a]	annual
critical load	250–1500 eq/ha/year[b]	annual
NO_x: critical level	30 μg/m^3	annual
critical load	5–35 kg N/ha/year[b]	annual
Ozone: critical level	0.2–10 ppm·h[a, c]	5 days–6 months

[a] Depending on the type of vegetation (see Part III).

[b] Depending on the type of soil and ecosystem (see Part III).

[c] AOT: Accumulated exposure Over a Threshold of 40 ppb.

pollutants that may adversely affect the ecosystem, and that the effects considered are only part of the spectrum of ecological effects. Effects on aquatic ecosystems were not evaluated, nor were effects on animals taken into account. Nevertheless, the available information indicates the importance of these pollutants and of their effects on terrestrial vegetation in the European Region.

CHAPTER 4

Use of the guidelines in protecting public health

When strategies to protect public health are under consideration, the air quality guidelines need to be placed in the perspective of total chemical exposure. The interaction of humans and the biosphere is complex. Individuals can be exposed briefly or throughout their lifetime to chemicals in air, water and food; exposures may be environmental and occupational. In addition, individuals vary widely in their response to exposure to chemicals; each person has a pre-existing status (for example, age, sex, pregnancy, pulmonary disease, cardiovascular disease, genetic make-up) and a lifestyle, in which such factors as exercise and nutrition play key roles. All these different elements may influence a person's susceptibility to chemicals. Various sensitivities also exist within the plant kingdom and need to be considered in protecting the environment.

The primary aim of these guidelines is to provide a uniform basis for the protection of public health and of ecosystems from adverse effects of air pollution, and to eliminate or reduce to a minimum exposure to those pollutants that are known or are likely to be hazardous. The guidelines are based on the scientific knowledge available at the time of their development. They have the character of recommendations, and it is not intended or recommended that they simply be adopted as standards. Nevertheless, countries may wish to transform the recommended guidelines into legally enforceable standards, and this chapter discusses ways in which this may be done. It is based on the report of a WHO working group (*1*). The discussion is limited to ambient air and does not include the setting of emission standards.

In the process of moving from a "guideline" or a "guideline value" to a "standard", a number of factors beyond the exposure–response relationship need to be taken into account. These factors include current concentrations of pollutants and exposure levels of a population, the specific mixture of air pollutants, and the specific social, economic and cultural conditions encountered. In addition, the standard-setting procedure may be influenced by the likelihood of implementing the standard. These considerations may lead to a standard above or below the respective guideline value.

DEFINITIONS

Several terms are in use to describe the tools available to manage ambient air pollution. To avoid confusion, definitions are needed for the terms used here – guideline, guideline value and standard – within this specific context.

Guideline

A guideline is defined as any kind of recommendation or guidance on the protection of human beings or receptors in the environment from adverse effects of air pollutants. As such, a guideline is not restricted to a numerical value but might also be expressed in a different way, for example as exposure–response information or as a unit risk estimate.

Guideline value

A guideline value is a particular form of guideline. It has a numerical value expressed either as a concentration in ambient air or as a deposition level, which is linked to an averaging time. In the case of human health, the guideline value provides a concentration below which no adverse effects or (in the case of odorous compounds), no nuisance or indirect health significance are expected, although it does not guarantee the absolute exclusion of effects at concentrations below the given value.

Standard

A standard is considered to be the level of an air pollutant, such as a concentration or a deposition level, that is adopted by a regulatory authority as enforceable. Unlike the case of a guideline value, a number of elements in addition to the effect-based level and the averaging time have to be specified in the formulation of a standard. These elements include:

- the measurement strategy
- the data handling procedures
- the statistics used to derive the value to be compared with the standard.

The numerical value of a standard may also include the permitted number of exceedings.

MOVING FROM GUIDELINES TO STANDARDS

The regulatory approach to controlling air pollution differs from country to country. Different countries have different political, regulatory and administrative approaches, and legislative and executive activities can be carried out at various levels such as national, regional and local. Fully effective air quality management requires a framework that guarantees a consistent derivation of air quality standards and provides a transparent basis for decisions with regard to risk-reducing measures and abatement strategies. In establishing such a framework, several issues should be considered, such as legal aspects, the protection of specific populations at risk, the role of stakeholders in the process, cost–benefit analysis, and control and enforcement measures.

Legal aspects

In setting air quality standards at the national or supranational level, a legislative framework usually provides the basis for the evaluation and decision-making process. The setting of standards strongly depends on the type of risk management strategy adopted. Such a strategy is influenced by country-specific sociopolitical considerations and/or supranational agreements.

Legislation and the format of air quality standards vary from country to country, but in general the following issues should be considered:

- identification and selection of pollutants to which the legislative instrument will apply;
- the process for making decisions about the appropriate standards;
- the numerical value of the standards for the various pollutants, applicable detection methods and monitoring methodology;
- actions to be taken to implement the standard, such as the definition of the time frame needed/allowed for achieving compliance with the standard, considering emission control measures and necessary abatement strategies; and
- identification of responsible enforcement authorities.

Depending on their position within a legislative framework, standards may or may not be legally binding. In some countries the national constitution contains provisions for the protection of public health and the environment. In general, the development of a legal framework on the basis of constitutional provisions comprises two regulatory actions. The first is the enactment of a formal legal instrument, such as an act, a law, an ordinance or a decree, and the second is the development of regulations, by-laws, rules and orders.

Air quality standards may be based solely on scientific and technical data on public health and environmental effects, but other aspects such as cost–benefit or cost–effectiveness may be also taken into consideration. In practice, there are generally several opportunities within a legal framework to address these economic aspects as well as other issues, such as technical feasibility, structural measures and sociopolitical considerations. These can be taken into account during the standard-setting procedure or at the level of designing appropriate measures to control emissions. This rather complicated process might result in several standards being set, such as an effect-oriented standard as a long-term goal and less stringent interim standards to be achieved within shorter periods of time.

Standards also depend on political choices as to which receptors in the environment should be protected and to what extent. Some countries have separate standards for the protection of public health and the environment. Moreover, the stringency of a standard can be influenced by provisions designed to take account of higher sensitivities of specific receptor groups, such as young children, sick or elderly people, or pregnant women. It might also be important to specify whether effects are considered for individual pollutants or for a combined exposure to several pollutants.

Air quality standards can set the reference point for emission control and abatement strategies on a national level. It should be recognized, however, that exposure to some pollutants is the result of long-range transboundary transport. In these cases adequate protection measures can only be achieved by appropriate international agreements.

Air quality standards should be regularly reviewed, and need to be revised as new scientific evidence on effects on public health and the environment emerges.

Standards often strongly influence the implementation of an air pollution control policy. In many countries, the exceeding of standards is linked to an obligation to develop action plans at the local, regional or national level to reduce air pollution levels. Such plans often address several pollution sources. Standards also play a role in environmental impact assessment procedures and in the provision of public information on the state of the environment. Provisions for such activities can be found in many national legal instruments.

Within national or supranational legislative procedures, the role of stakeholders in the process of standard-setting also needs to be considered. This is dealt with in more detail below.

Items to be considered in setting standards

Within established legal frameworks and using air quality guidelines as a starting point, development of standards involves consideration of a number of issues. These are in part determined by characteristics of populations or physical properties of the environment. A number of these issues are discussed below.

Adverse health effects

In setting a standard for the control of an environmental pollutant, the effects that the population is to be protected against need to be defined. A hierarchy of effects on health can be identified, ranging from acute illness and death through chronic and lingering diseases and minor and temporary ailments, to temporary physiological or psychological changes. The distinction between adverse and non-adverse effects poses considerable difficulties. Of course, more serious effects are generally accepted as adverse. As one considers effects that are either temporary and reversible, or involve biochemical or functional changes whose clinical significance is uncertain, judgements must be made as to which of these less serious effects should be considered adverse. With any definition of adversity, a significant degree of subjectivity and uncertainty remains. Judgements as to adversity may differ between countries because of factors such as different cultural backgrounds and different background levels of health status.

In some cases, the use of biomarkers or other indicators of exposure may provide a basis for standard-setting. Changes in such indicators, while not necessarily being adverse in themselves, may be predictors of significant effects on health. For example, the blood lead concentration can provide information on the likelihood of impairment of neurobehavioural development.

Special populations at risk

Sensitive populations or groups are defined here as those impaired by concurrent disease or other physiological limitations, and those with specific characteristics that make the health consequences of exposure more significant (such as the developmental phase in children or reduction in reserve capacity in the elderly). In addition, other groups may be judged to be at special risk because of their exposure patterns or due to an increased effective dose for a given exposure. Sensitive populations may vary from country to country owing to differences in the number of people lacking access to adequate medical care, in the existence of endemic disease, in the prevailing genetic factors, or in the prevalence of debilitating diseases, nutritional deficiencies or lifestyle factors. It is up to the politician to decide which specific groups at risk should be protected by the standards (and thus which should not be protected).

Exposure–response relationships

A key factor to be considered in developing standards is information about the exposure–response relationship for the pollutant concerned. For a number of pollutants an attempt has been made to provide exposure–response relationships in the revised version of the guidelines. For particulate matter and ozone, detailed tables specifying the exposure–response relationship are provided. The information included in these tables is derived from epidemiological studies of the effects of these pollutants on health. Such information is available for only a few of the pollutants considered in the guidelines. For known "no-threshold compounds" such as the carcinogen benzene, quantitative risk assessment methods provide estimates of risk at different exposure concentrations.

When developing standards, regulators should consider the degree of uncertainty about exposure–response relationships provided in the guidelines. Differences in the population structure, climate and geography that can have an impact on the prevalence, frequency and severity of effects may modify the exposure–response relationships provided in the guidelines.

Exposure characterization

An important factor to be considered in developing standards is that of how many people are exposed to concentrations of concern and the distribution of exposure among various population groups. Current distributions of exposure should be considered, together with those that are likely to occur should the standard be met. Besides using monitoring data, results of exposure modelling can be used at this stage. The origins of pollutants, including long-range transport and its contribution to ambient levels, should also be evaluated.

The extent to which ambient air quality estimates from monitoring networks or models correspond to personal exposure in the population is also a factor to be considered in the standard-setting. This will depend on the pollutant in question (for example, personal exposure to carbon monoxide is poorly characterized by fixed-site monitors) as well as on a number of local characteristics, including lifestyle, climatic conditions, spatial distribution of pollution sources and local determinants of pollution dispersion.

Other important exposure-related concerns include:

- how much of total human exposure is due to ambient, outdoor sources as opposed to indoor sources; and

- where multiple routes of exposure are important, how to apportion the regulatory burden among the different routes of exposure (such as lead from air sources versus lead from paint, water pipes, etc.).

These factors may vary substantially across countries. For example, indoor air pollution levels might be quite substantial in countries in which fossil and/or biomass fuels are used in homes.

Risk assessment

In general, the central question in developing air quality standards to protect public health or ecosystems is the degree of protection associated with different pollution levels at which standards might be established. In the framework of quantitative risk assessment, various proposals for standards can be considered in health or ecological risk models. These models provide a tool that is increasingly used to inform decision-makers about some of the possible consequences associated with various options for standards, or the reduction in adverse effects associated with moving from the current situation to a particular standard.

The first two steps in risk assessment, namely hazard identification and, in some cases, development of exposure–response relationships, have been provided in these guidelines and are discussed in greater detail in later chapters. The third step, exposure analysis, predicts changes in exposure associated with reductions in emissions from a specific source or groups of sources under different control scenarios. Instead of exposure estimates, ambient concentrations (based on monitoring or modelling) are often used as the inputs to a risk assessment. This is in part because of the availability of information on concentration–response relationships from epidemiological studies in which fixed-site monitors were used.

The final step in a regulatory risk assessment is the risk characterization stage, whereby exposure estimates are combined with exposure–response relationships to generate quantitative estimates of risk (such as how many individuals may be affected). Regulatory risk assessments are likely to result in different risk estimates across countries, owing to differences in exposure patterns and in the size and characteristics of sensitive populations and those at special risk.

It is important to recognize that there are many uncertainties at each stage of a regulatory risk assessment. The results of sensitivity and uncertainty analyses should be presented so as to characterize the impact of major uncertainties on the risk estimates. In addition, the methods used to conduct the risk assessments should be clearly described and the limitations and caveats associated with the analysis should be discussed.

Acceptability of risk

The role of a regulatory risk assessment in developing standards may differ from country to country, owing to differences in the legal framework and availability of information. Also, the degree of acceptability of risk may vary between countries because of differences in social norms, degree of adversity and risk perception among the general population and various stakeholders. How the risks associated with air pollution compare with those from other pollution sources or human activities may also influence risk acceptability.

In the absence of clearly identified thresholds for health effects for some pollutants, the selection of a standard that provides adequate protection of public health requires an exercise of informed judgement by the regulator. The acceptability of the risks and, therefore, the standard selected will depend on the effect, on the expected incidence and severity of the potential effects, on the size of the population at risk, and on the degree of scientific certainty that the effects will occur at any given level of pollution. For example, if a suspected health effect is severe and the size of the population at risk is large, a more cautious approach would be appropriate than if the effect were less troubling or if the exposed population were small.

Cost–benefit analysis

Two comprehensive techniques provide a framework for comparing monetarized costs and benefits of implementing legislation or policy: cost–effectiveness analysis and cost–benefit analysis. These two techniques differ in their treatment of benefits. In cost–benefit analysis, costs and benefits (for example, avoided harm, injury or damage) of implemented control measures are compared using monetary values. In cost–effectiveness analysis, the costs of control measures are reported in quantitative terms, such as cost per ton of pollutant or cost per exposure unit. That is, the benefits are described in their own physical, chemical or biological terms, such as reduced concentrations or emissions, or avoided cases of illness, crop losses or damage to ecosystems.

Analysis of control measures to reduce ambient pollutant levels

Control measures to reduce emissions of many air pollutants are known. Direct control measures at the source are readily expressed in monetary terms. Indirect control measures, such as alternative traffic plans or changes in public behaviour, may not all be measurable in monetary terms but their impact should be understood. Effective control measures should be designed to deal with secondary as well as primary pollutants.

Cost identification should include costs of investment, operation and maintenance, both for the present and for the future. Unforeseen effects, technical innovations and developments, and indirect costs arising during implementation of the regulation are additional complicating factors. Cost estimates derived in one geographical area may not be generally transferable to other areas.

Air quality assessment has to provide information about expected air quality, both with and without implementation of control measures. Typically, the assessment will be based on a combination of air quality monitoring data and dispersion modelling. These two assessment methods are complementary, and must be seen as equally important inputs to the assessment process.

For the assessment, several types of data have to be acquired:

- measured concentrations for relevant averaging times (hourly, daily, seasonal) with information on site classification;
- emission data from all significant sources, including emission conditions (such as stack height) and with sufficient information on spatial and temporal variation; and
- meteorological and topographical data relevant to dispersion of the emissions.

Defining the scope and quantifying the benefits

The air quality guidelines are based on health and ecosystem endpoints determined by consensus. This does not imply that other effects on health and the ecosystem that were not considered in the guidelines may not occur or are unimportant. After assessing the local situation, other health- and ecosystem-related benefit categories should be considered in the analysis.

It is a difficult and comprehensive task to quantify the benefit categories included in a cost–benefit analysis. Some indicators of morbidity, such as the use of medication, the number of hospital admissions or work days lost, can be quantified. Other effects, such as premature death or excess mortality, present more difficult problems. Wellbeing, the quality of life or the value of ecosystems may be very difficult to express in monetary terms. In different countries, values assigned to benefit categories might differ substantially owing to different cultural attitudes. Despite these uncertainties, it is better to include as many of the relevant benefit categories as possible, even if the economic assessment is uncertain or ambiguous. A clear understanding of the way in which the economic assessment has been undertaken is important and should be reported.

Comparison of benefits with and without control actions

This step involves combining the information on exposure–response relationships with that on air quality assessment, and applying the combined information to the population at risk. Additional data needed in this step include specification of the population at risk, and determination of the prevalence of the different health effects in the population at risk.

Comparison of costs and benefits

Monetary valuation of control actions and of health and environmental effects may be different in concept and vary substantially from country to country. In addition to variations in assessing costs, the relative value of benefit categories, such as benefits to health or building materials, will vary. Thus, the result of comparing costs and benefits in two areas with otherwise similar conditions may differ significantly.

The measures taken to reduce one pollutant may increase or decrease the concentration of other pollutants. These additional effects should be considered, even if they result from exposure to pollutants not under consideration in the primary analysis. Pollutant interactions pose additional complications. Interaction effects may possibly lead to double counting of costs, or to disregarding some costly but necessary action. The same argumentation can be used when estimating benefits.

Sensitivity and uncertainty analysis

Sensitivity analysis includes comparisons of the results of a particular cost–benefit analysis with that of other studies, recalculation of the whole chain of the analysis using other assumptions, or the use of ranges of values. Specifically, a range of values may be used, such as for value of statistical life. Knowledge of the costs of control measures tends to be better developed than that of the benefits to health and ecosystems, and thus costs tend to be more accurately estimated than benefits. In addition, costs tend to be overestimated and benefits underestimated. One important reason for underestimating the benefits is not considering some important benefit categories because of lack of information. Another reason is the variability of the databases available for monetary assessment of benefits.

Many uncertainties are connected with the steps of cost–benefit and cost-effectiveness analysis, such as exposure, exposure–response, control cost estimates and benefits valuation. The results of sensitivity and uncertainty analyses should be presented so as to characterize the impact of major uncertainties on the result of the analysis. In addition, the methods used to conduct the analysis should be clearly described, and the limitations and caveats associated with the analysis should be discussed. Transparency of the analysis is most important.

Involvement of stakeholders and public awareness

The development of standards should encompass a process involving stakeholders that ensures, as much as possible, social equity or fairness to all involved parties. It should also provide sufficient information to guarantee understanding by stakeholders of the scientific and economic consequences. A review by stakeholders of the standard-setting process, initiated at an early stage, is helpful. Transparency in the process of moving from air quality guidelines to standards helps the public to accept necessary measures.

The participation of all those affected by the procedure of standard-setting – industry, local authorities, nongovernmental organizations and the general public – at an early stage of standard derivation is strongly recommended. If these parties are involved in the process at an early stage their cooperation is more likely to be elicited.

Raising public awareness of the health and environmental effects of air pollution is also an important means to obtain public support for necessary control actions, such as with respect to vehicle emissions. Information about the quality of air (such as warnings of air pollution episodes) and the entailed risks (risk communication) should be published in the media to keep the public informed.

IMPLEMENTATION

The main objectives of the implementation of air quality standards are: (*a*) to define the measures needed to achieve the standards; and (*b*) to establish a suitable regulatory strategy and legislative instrument to achieve this goal. Long- as well as medium-term goals are likely to be needed.

The implementation process should ensure a mechanism for regular assessment of air quality, set up the abatement strategies, and establish the enforcement regulations. Also, the impact of control actions should be assessed, both for public health and environmental effects, for example by the use of epidemiological studies and integrated ecosystems monitoring. Epidemiological studies of the effects of air pollutants on health should be repeated as control measures are implemented. Changes to the mixture of air pollutants and in the composition of complex pollutants such as particulate matter may occur, and changes in exposure–response relationships should be expected.

Assessment of air quality

Air quality assessment has an important role to play within the implementation of an air quality management strategy. The goals of air quality assessment are to provide the air quality management process with relevant data through a proper characterization of the air pollution situation, using monitoring and/or modelling programs and projection of future air quality associated with alternative strategies. Dispersion models can be used very effectively in the design of the definitive monitoring network.

Monitoring methods

The monitoring method (automatic, semi-automatic or manual) adopted for each pollutant should be a standard or reference method, or be validated against such methods. The full description of the method should include information on the sampling and analytical method, on the quality assurance and quality control (internal and external) procedures and on the methods of data management, including data treatment, statistical handling of the data and data validation procedures.

Quality assurance/quality control procedures are an essential part of the measurement system, the aim being to reduce and minimize errors in both the instruments and management of the networks. These procedures should ensure that air quality measurements are consistent (and can be used to give a reliable assessment of ambient air quality) and harmonized over a scale as large as possible, especially in the area of the implementation of the standard.

Design of the monitoring network

An air quality monitoring network can consist of fixed and/or mobile monitoring stations. Although such a network is a fundamental tool for any air quality assessment, its limitations should be borne in mind.

In designing a monitoring network, a primary requirement is to have information about emissions from the dominant and/or most important sources of pollutants. Second, a pilot (or screening) study is needed to gain a good understanding of the geographical distribution of pollutants and to identify the areas with the highest concentrations. Such a screening study can be performed using dispersion models, with the emission inventory as input, combined with a monitoring study using inexpensive passive samplers in a rather dense network.

The selection strategy for site locations generally varies for different pollutants. The number and distribution of sampling sites required in any network depend on the area to be covered, the spatial variability of the emissions being measured, and the purpose for which the data should be used. Meteorological and topographical conditions as well as the density, type and strength of sources (mobile and stationary) must be considered.

Different types of monitoring station are likely to be needed to provide data at a regional or local level. In monitoring rural and urban areas, specific attention should be paid to sites affected by defined sources such as traffic and other "hot-spots". The representativeness of each site should be defined and assessed. Micro-scale conditions, including the buildings around the stations (street canyons), traffic intensity, the height of the sampling point, distances to obstacles, and the effects of the local sources must be kept in mind.

Air quality modelling

Air quality models are used to establish a relationship between emissions and air quality in a given area, such as a city or region. On the basis of emission data, of atmospheric chemistry, and of meteorological, topographical and geographical parameters, modelling gives an opportunity (*a*) to calculate the projected concentration or deposition of the pollutants in regions, and (*b*) to predict the air pollution level in those areas where air sampling is not performed. Measured concentrations should be used for evaluating and validating models, or even as input data. These measurements improve the accuracy of the concentrations calculated by models by allowing refinement and development of the modelling strategies adopted.

Abatement strategies

Abatement strategies are the set of measures to be taken to reduce pollutant emissions and therefore to improve air quality. Authorities should consider the measures necessary in order to meet the standards. An important factor in selecting abatement strategies is deciding the geographical scale of the area(s) that are considered not to meet the standard(s) and the geographical scale of the area for which control should be applied. In defining the geographical scale for abatement strategies, the extent of the transport of pollution from neighbouring areas should be considered. This may involve action at supranational, national, regional or local levels.

Supranational, national, regional and local actions form a hierarchy of approaches. Action at the supranational or national level is likely to be most effective in reducing background levels of air pollution. Local air quality management measures may be needed to address specific local problems, and such measures may need to be implemented urgently to deal with special pollution problems. National and supranational plans should specify the extent of the reduction in levels of air pollution that is required and the time-scale for achieving that reduction.

In addition to the comprehensive programme of emission control designed to reduce average pollution levels and the risk of high pollution episodes, short-term actions may be required for the period when the pollution episodes may occur. Such actions, however, should be considered to be applicable in a transitional period only or as a contingency plan. The objective of measures applied on a larger scale is to minimize the occurrence of local air pollution episodes. A link between control of emissions and ambient air quality is required and may need to be demonstrated. Emission-based air quality standards represent one possible step in this process.

Enforcement

The government of each country establishes the responsibilities for implementing air quality standards. Responsibilities for overseeing different aspects of compliance can be distributed among national, regional and local governments depending on the level at which it is necessary to take action.

Success in the enforcement of standards is influenced by the technology applied and the availability of financial resources to industry and government. Compliance with standards may be ensured by various approaches such as administrative penalties or economic incentives. Sufficient staff and other resources are needed to implement the policy actions effectively.

Periodic reports on compliance and trends in pollutant emissions and concentrations should be developed and disseminated to the public. These reports should also predict trends. It is important that the public be aware of the importance of meteorological factors in controlling pollution levels, as these may produce episodes of pollution that are not within the control of the regulatory authorities.

REFERENCE

1. *Guidance for setting air quality standards. Report on a WHO Working Group*. Copenhagen, WHO Regional Office for Europe, 1998 (document EUR/ICP/EHPM 02 01 02).

PART II

EVALUATION OF RISKS TO HUMAN HEALTH

CHAPTER 5

Organic pollutants

5.1 Acrylonitrile

Exposure evaluation

On the basis of large-scale calculations using dispersion models, the average annual ambient air concentration of acrylonitrile in the Netherlands was estimated to be about 0.01 μg/m³ *(1),* which is below the present detection limit of 0.3 μg/m³ *(1, 2).* Production figures *(1)* indicate that, in 8 out of 10 European countries for which data are available, ambient concentrations of acrylonitrile are lower or markedly lower than this. Near industrial sites, air concentrations can exceed 100 μg/m³ over a 24 hour period, but are usually less than 10 μg/m³ at a distance of about 1 km. Acrylonitrile concentrations in the air at the workplace have exceeded 100 mg/m³, but shift averages are usually in the range of 1–10 mg/m³. Exposure from smoking is possible if acrylonitrile is used for tobacco fumigation, and could amount to 20–40 μg daily for an average smoker.

A more sensitive method of determination, with a detection limit below 0.1 μg/m³, is required in order to examine concentrations in the ambient air and to allow populations at possible risk to be identified.

Health risk evaluation

Acute and noncancer chronic toxicity may occur at concentrations still reported in some industries. Subjective complaints were reported in acute exposure to 35 mg/m³, and in chronic exposure to 11 mg/m³, 4.2–7.2 mg/m³ or 0.6–6 mg/m³. Teratogenic effects in animals were observed at 174 mg/m³ and carcinogenicity was shown in rats exposed for 2 years to 44 mg/m³.

Twelve epidemiological studies investigating the relationship between acrylonitrile exposure and cancer are available; only five indicate a carcinogenic risk from exposure to acrylonitrile *(1).* Negative studies suffered from small cohort size, insufficient characterization of exposure, short follow-up times and relatively young cohorts. Although four of the remaining five epidemiological studies indicate a higher risk of lung cancer, and one study showed a higher mortality rate for liver, gall bladder and cystic duct cancer, all have problems with regard to methodology, definition and/or size of the population, existence of exposure to other carcinogens, and duration of the follow-up period.

In laboratory animals an increased incidence of tumours of the central nervous system, Zymbal gland, stomach, tongue, small intestine and mammary glands was observed at all doses tested *(3)*. There is nevertheless a clear difference between animal and human studies concerning the tumorigenic response to acrylonitrile: no lung tumours have been produced in animals and no brain tumours have been observed in humans.

Acrylonitrile was placed in IARC Group 2A *(3)* on the basis of sufficient evidence of its carcinogenicity in experimental animals and limited evidence of its carcinogenicity in humans.

The epidemiological study by O'Berg *(4)* presents the clearest available evidence of acrylonitrile as a human lung carcinogen. Furthermore, in this study there were no confounding exposures to other carcinogenic chemicals during exposure to acrylonitrile. It was therefore used to make an estimate of the incremental unit risk. As this study has now been updated to the end of 1983 for cancer incidence and to the end of 1981 for overall mortality, the most recent data are used here *(5)*. Of 1345 workers exposed to acrylonitrile, a total of 43 cases of cancer occurred versus 37.1 expected. Ten cases of lung cancer were observed versus 7.2 expected, based on the company rates. Lung cancer, which had been the focus of the previous report *(4)*, remained in excess but not as high as before; 2 new cases occurred after 1976, with 2.8 expected. This means that the relative risk (RR) would be 10/7.2 = 1.4, significantly lower than in the previous report. On the assumption made by the US Environmental Protection Agency *(6)* for the first O'Berg study *(4)* that the 8-hour time-weighted average exposure was 33 mg/m^3 (15 ppm), and with an estimated work duration of 9 years, the average lifetime daily exposure (X) is estimated to be 930 µg/m^3 ($X = 33\ mg/m^3 \times 8/24 \times 240/365 \times 9/70$).

Using the average relative risk model, the lifetime unit risk (UR) for exposure to 1 µg/m^3 can be calculated to be 1.7×10^{-5} [$UR = P_o(RR - 1)/X = 0.04(1.4 - 1)/930$].

Using animal data, an upper-bound risk of cancer associated with a lifetime inhalation exposure to acrylonitrile was calculated from a rat inhalation study *(7)* to be 1.5×10^{-5} *(6)*.

The calculated unit risk based on the human study is consistent with that of the animal study, although the human estimate is uncertain, particularly because of the lack of documentation on exposure.

Guidelines

Because acrylonitrile is carcinogenic in animals and there is limited evidence of its carcinogenicity in humans, it is treated as if it were a human carcinogen. No safe level can therefore be recommended. At an air concentration of 1 μg/m^3, the lifetime risk is estimated to be 2×10^{-5}.

References

1. *Criteriadocument over acrylonitril* [Acrylonitrile criteria document]. The Hague, Ministry of Housing, Spatial Planning and Environment, 1984 (Publikatiereeks Lucht, No. 29).
2. GOING, J.E. ET AL. *Environmental monitoring near industrial sites: acrylonitrile.* Washington, DC, US Environmental Protection Agency, 1979 (Report No. EPA-560/6-79-003).
3. *Chemicals, industrial processes and industries associated with cancer in humans. IARC Monographs, Volumes 1 to 29.* Lyons, International Agency for Research on Cancer, 1982 (IARC Monographs on the Evaluation of the Carcinogenic Risk of Chemicals to Humans, Supplement 4), pp. 25–27.
4. O'BERG, M.T. Epidemiologic study of workers exposed to acrylonitrile. *Journal of occupational medicine,* **22**: 245–252 (1980).
5. O'BERG, M.T. ET AL. Epidemiologic study of workers exposed to acrylonitrile: an update. *Journal of occupational medicine,* **27**: 835–840 (1985).
6. *Health assessment document for acrylonitrile.* Washington, DC, US Environmental Protection Agency, 1983 (Final report EPA-600/8-82-007F).
7. QUAST, J.F. ET AL. *A two-year toxicity and oncogenicity study with acrylonitrile following inhalation exposure of rats. Final report.* Midland, MI, Dow Toxicology Research Laboratory, 1980.

5.2 Benzene

Exposure evaluation

Sources of benzene in ambient air include cigarette smoke, combustion and evaporation of benzene-containing petrol (up to 5% benzene), petrochemical industries, and combustion processes.

Mean ambient air concentrations of benzene in rural and urban areas are about 1 μg/m^3 and 5–20 μg/m^3, respectively. Indoor and outdoor air levels are higher near such sources of benzene emission as filling stations.

Inhalation is the dominant pathway for benzene exposure in humans. Smoking is a large source of personal exposure, while high short-term exposures can occur during refuelling of motor vehicles. Extended travel in motor vehicles with elevated air benzene levels (from combustion and evaporative emissions) produces exposures reported from various countries that are second only to smoking as contributors to the intensity of overall exposure. The contribution of this source to cumulative ambient benzene exposure and associated cancer risk comprises about 30% when the travel time is one hour, a duration not untypical for urban and suburban commuting by the general population.

Health risk evaluation

The most significant adverse effects from prolonged exposure to benzene are haematotoxicity, genotoxicity and carcinogenicity.

Chronic benzene exposure can result in bone marrow depression expressed as leukopenia, anaemia and/or thrombocytopenia, leading to pancytopenia and aplastic anaemia. Decreases in haematological cell counts and in bone marrow cellularity have been demonstrated in mice after inhalation of concentrations as low as 32 mg/m^3 for 25 weeks. Rats are less sensitive than mice. In humans, haematological effects of varying severity have occurred in workers occupationally exposed to high levels of benzene. Decreased red and white blood cell counts have been reported above median levels of approximately 120 mg/m^3, but not at 0.03–4.5 mg/m^3. Below 32 mg/m^3, there is only weak evidence of effects.

The genotoxicity of benzene has been extensively studied. Benzene does not induce gene mutations in *in vitro* systems, but several studies have demonstrated induction of both numerical and structural chromosomal aberrations, sister chromatid exchanges and micronuclei in experimental animals and humans after *in vivo* benzene exposure. Some studies in humans have demonstrated chromosomal effects at mean workplace exposures as low as 4–7 mg/m^3. The *in vivo* data indicate that benzene is mutagenic.

The carcinogenicity of benzene has been established both in humans and in laboratory animals. An increased mortality from leukaemia has been demonstrated in workers occupationally exposed. Several types of tumour, primarily of epithelial origin, have been induced in mice and rats after oral exposure and inhalation exposure at 320–960 mg/m^3; these include tumours in the Zymbal gland, liver, mammary gland and nasal cavity. Lymphomas/leukaemias have also been observed, but with lower frequency. The results indicate that benzene is a multisite carcinogen.

Because benzene is characterized as a genotoxic carcinogen and recent data gathered in humans and mice suggest mutagenic potential *in vivo*, establishment of exposure duration and concentration in the human exposure studies is of major importance for the calculation of cancer risk estimates. The Pliofilm cohort is the most thoroughly studied. It was noted that significant exposures to other substances at the studied facilities were probably not a complicating factor, but that exposure estimates for this cohort vary considerably. Three different exposure matrices have been used to describe the Pliofilm cohort, i.e. those reported by Crump & Allen *(1)*, by Rinsky et al. *(2)*, and a newer and more extensive one by Paustenbach et al. *(3)*. The main difference between the first two is that the exposure estimates by Crump & Allen are greater for the early years, during the 1940s. Paustenbach et al. have, among other things, considered short-term, high-level exposure, background concentrations and absorption through the skin, which leads to exposure levels 3–5 times higher than those calculated by Rinsky et al. Compared to the Crump & Allen estimates, Paustenbach et al. arrived at higher exposure estimates for some job classifications, and lower ones for some others.

Within the most recently updated Pliofilm cohort, Paxton et al. *(4, 5)* conducted an extended regression analysis with exposure description for the 15 leukaemia cases and 650 controls. They used all three exposure matrices, which gave estimates of 0.26–1.3 excess cancer cases among 1000 workers at a benzene exposure of 3.2 mg/m^3 (1 ppm) for 40 years (Table 8).

Crump *(7)* calculated unit risk estimates for benzene using the most recently updated data for the Pliofilm cohort and a variety of models

Table 8. Published leukaemia risk estimates for the Plioform cohort at two benzene exposure levels

Cases per 1000 workers exposed to:			
3.2 mg/m³ (1 ppm)	**0.32 mg/m³ (0.1 ppm)**	**Exposure matrix**	**Reference**
5.3	–	Rinsky et al. *(2)*	Brett et al. *(6)*
0.5–1.6	–	Rinsky et al. *(2)*	
		Crump & Allen *(1)*	Brett et al. *(6)*
1.3	0.12	Rinsky et al. *(2)*	Paxton et al. *(4, 5)*
0.26	0.026	Crump & Allen *(1)*	Paxton et al. *(4, 5)*
0.49	0.048	Paustenbach et al. *(3)*	Paxton et al. *(4, 5)*

(Table 9). Multiplicative risk models were found to describe the cohort data better than additive risk models and cumulative exposure better than weighted exposures. Dose–responses were essentially linear when the Crump & Allen exposure matrix was used but, according to the author, there was evidence of concentration-dependent nonlinearity in dose–responses derived using the Paustenbach et al. exposure matrix. In that case, the best-fitting model was quadratic.

As can be seen in Table 9, the concentration-dependent model gives a much lower risk estimate than the other models when the Paustenbach et al. exposure matrix is used. In such a model, the concentration of benzene is raised to the second power and thus given greater weight than the duration of exposure. Although there are biological arguments to support the use of a concentration-dependent model, many of the essential data are preliminary and need to be further developed and peer reviewed.

Models giving equal weight to concentration and duration of exposure have been preferred here for the derivation of a risk estimate. Using multiplicative risk estimates and a cumulative exposure model, Crump *(7)* calculated a unit risk for lifetime exposure of $1.4–1.5 \times 10^{-5}$ per ppb with the Paustenbach et al. exposure matrix, and of 2.4×10^{-5} per ppb with the Crump & Allen exposure matrix. If expressed in µg/m³, the unit risk would thus range from 4.4×10^{-6} to 7.5×10^{-6}. With an additive model instead of a multiplicative model, the risk estimate would have been somewhat smaller. If similar linear extrapolations were done on the occupational cancer risk

Table 9. Model-dependent worker risk and lifetime unit risk estimates for exposure to benzene for the Plioform cohort by Crump *(7)* [a]

Risk estimate	Linear	Nonlinear	Intensity dependent	Exposure reference
Cases per 1000 workers exposed to 3.2 mg/m^3 (1 ppm)	5.1	5.0	5.1	Crump & Allen *(1)*
	3.8	2.9	0.036	Paustenbach et al. *(3)*
Unit risk per ppb	2.4×10^{-5}	2.4×10^{-5}	2.4×10^{-5}	Crump & Allen *(1)*
	1.5×10^{-5}	1.4×10^{-5}	1.7×10^{-10}	Paustenbach et al. *(3)*
Unit risk per µg/m^3 [b]	7.5×10^{-6}	7.5×10^{-6}	7.5×10^{-6}	Crump & Allen *(1)*
	4.7×10^{-6}	4.4×10^{-6}	5.3×10^{-11}	Paustenbach et al. *(3)*

[a] Multiplicative risk model, cumulative exposure.

[b] Calculated by converting ppb to µg/m^3.

estimates by Paxton et al. (Table 8), unit risks lower by up to about one order of magnitude would result.

Guidelines

Benzene is carcinogenic to humans and no safe level of exposure can be recommended. For purposes of guideline derivation, it was decided to use the 1994 risk calculation of Crump rather than to derive new estimates. It was recognized that this use of existing analyses of the most recently updated cohort ruled out the inclusion of certain of the analyses noted earlier.

The geometric mean of the range of estimates of the excess lifetime risk of leukaemia at an air concentration of 1 µg/m^3 is 6×10^{-6}. The concentrations of airborne benzene associated with an excess lifetime risk of 1/10 000, 1/100 000 and 1/1 000 000 are, respectively, 17, 1.7 and 0.17 µg/m^3.

References

1. CRUMP, K. & ALLEN, B. *Quantitative estimates of risk of leukemia from occupational exposure to benzene*. Washington, DC, US Department of Labor, 1984 (OSHA Docket H-059b, Exhibit 152, Annex B).
2. RINSKY, R.A. ET AL. Benzene and leukaemia. An epidemiologic risk assessment. *New England journal of medicine*, **316**: 1044–1050 (1987).
3. PAUSTENBACH, D.J. ET AL. Reevaluation of benzene exposure for the pliofilm (rubberworker) cohort (1936–1976). *Journal of toxicology and environmental health*, **36**: 177–231 (1992).

4. PAXTON, M.B. ET AL. Leukaemia risk associated with benzene exposure in the pliofilm cohort: I. Mortality update and exposure distribution. *Risk analysis,* **14**: 147–154 (1994).
5. PAXTON, M.B. ET AL. Leukaemia risk associated with benzene exposure in the pliofilm cohort. II. Risk estimates. *Risk analysis,* **14**: 155–161 (1994).
6. BRETT, S.M. ET AL. Review and update of leukaemia risk potentially associated with occupational exposure to benzene. *Environmental health perspectives*, **82**: 267–281 (1989).
7. CRUMP, K.S. Risk of benzene-induced leukaemia: a sensitivity analysis of the pliofilm cohort with additional follow-up and new exposure estimates. *Journal of toxicology and environmental health*, **42**: 219–242 (1994).

5.3 Butadiene

Exposure evaluation

In a survey of butadiene monomer, polymer and end-user industries in the United States, the geometric mean concentration for full-shift exposure for all job categories was 0.098 ppm and the arithmetic mean was 2.12 ppm *(1)*.

Although data for ambient air levels in Europe are limited, reported concentrations in urban air generally ranged from less than 2 µg/m^3 to 20 µg/m^3 *(2, 3)*. Mean levels in indoor air in a small number of Canadian homes and offices were 0.3 µg/m^3 *(4)*. Sidestream cigarette smoke contains 1,3-butadiene at approximately 0.4 mg/cigarette, and levels of butadiene in smoky indoor environments are typically 10–20 µg/m^3 *(5)*.

Health risk evaluation

Irritation or effects on the central nervous system may be associated with acute exposure to high concentrations of butadiene. Nevertheless, carcinogenicity is considered to be the critical effect for the derivation of air quality guidelines.

1,3-Butadiene has induced a wide variety of tumours in rats and mice, with mice being considerably more sensitive than rats. There are widely divergent points of view as to which animal species – the rat or the mouse – is most appropriate for use in human risk assessments for butadiene *(6, 7)*.

Epidemiological studies, while relatively few in number, suggest that there is equivocal evidence for an association between exposure to butadiene and lymphohaematopoietic cancer. In 1992, IARC classified butadiene in Group 2A (probably carcinogenic to humans). Preliminary (unpublished) reports suggest, however, that there may be an association between butadiene exposure and leukaemia in workers in the synthetic rubber industry.

The genotoxicity of butadiene has been studied in a variety of *in vitro* and *in vivo* mutagenicity assays, and the data overwhelmingly suggest that the induction of cancer requires the metabolism of butadiene to its DNA-reactive metabolites. Butadiene is mutagenic in both bacterial and mammalian systems. The butadiene metabolites epoxybutene and diepoxybutane are also carcinogenic and genotoxic *in vivo*. Butadiene is metabolized to epoxides to a significantly lesser extent in human tissues than in mice and rats. The differences between mice and rats observed *in vitro* are supported by *in vivo* studies, indicating that mice form very high levels of epoxides compared to rats when exposed to butadiene. In general, the data support the conclusion that the metabolism of butadiene in humans is more similar to that in rats, a relatively insensitive species to butadiene carcinogenicity, than to that in mice, a highly sensitive species. It should be recognized, however, that inter-individual differences in butadiene metabolism may exist that will influence the extent to which butadiene epoxides are formed.

In the only published human study, of 40 individuals occupationally exposed to butadiene at levels typical of an industrial setting (1–3 ppm), there were no significant increases in chromosome aberrations, micronucleus formation or sister chromatid exchanges in peripheral blood lymphocytes *(8)* compared to controls (30 individuals). This observation is of particular interest since butadiene concentrations as low as 6.25 ppm increased the occurrence of the same indicators of genetic damage in the bone marrow and peripheral blood lymphocytes of mice.

Several different risk assessments have been conducted for butadiene, and a number of these for occupational exposures to butadiene have been summarized by the US Occupational Safety and Health Administration *(9)*. The estimates in these risk assessments were based on different assumptions. Some were adjusted for absorbed dose, since changes in butadiene absorption will occur in animals with changes in the inhaled concentration *(10)*. For the most part, they were based on the multistage model. There was considerable variation in human cancer risk estimates depending on the animal species used for the calculations, with those based on tumour data in mice being 100–1000 times higher than those based on tumour data in rats.

Unit risk estimates for cancer associated with continuous lifetime exposure to butadiene in ambient air have been reported *(11–13)*. Values estimated by the Californian Air Resources Board in 1992 *(11)*, based on adjustment of dose for absorption *(10)* and tumour incidence in mice *(14)* and rats *(15)*, were 0.0098 and 0.8 per ppm, respectively. The value estimated by the US Environmental Protection Agency's Integrated Risk Information System (IRIS), based on linearized multistage modelling of data from an earlier, limited US National Toxicology Program (NTP) bioassay in mice, was 2.8×10^{-4} per $\mu g/m^3$ *(12)*. Values estimated by the National Institute of Public Health and Environmental Protection (RIVM) in the Netherlands, based on linearized multistage modelling of the incidence of lymphocytic lymphoma and haemangiosarcomas of the heart in mice in the most recent NTP bioassay *(14)*, were in the range $0.7–1.7 \times 10^{-5}$ per $\mu g/m^3$ *(13)*.

Estimates of human cancer risk could be improved by the inclusion of mechanistic information such as *in vivo* toxicokinetic data, genotoxicity data, and data from the recent epidemiology reassessment. For example, new data on levels of butadiene epoxides in blood and tissues in laboratory animals *(16–18)* could be used to replace the earlier absorption data *(10)*. Additionally, physiologically based pharmacokinetic models developed since earlier attempts to apply this approach to risk assessment have been greatly improved, most notably by the incorporation of model parameters that have been experimentally measured rather than empirically estimated. None the less, none of the models published to date incorporates the necessary information on the formation, removal and distribution of diepoxybutane.

Guidelines

Quantitative cancer risk estimates vary widely, in particular depending on the test species used. No definitive conclusions can yet be made as to which species should be used for risk estimates. New, as yet unpublished epidemiological data might have an impact on the risk estimates and hence on the derivation of a guideline value. In the light of these considerations, no guideline value can be recommended at this time.

References

1. FAJEN, J.M. ET AL. Industrial exposure to 1,3-butadiene in monomer, polymer, and end user industries. *In*: Sorsa, M. et al., ed. *Butadiene and styrene: assessment of health hazards.* Lyons, International Agency for Research on Cancer, 1993 (IARC Scientific Publications, No. 127), pp. 3–13.
2. NELIGAN, R.E. Hydrocarbons in the Los Angeles atmosphere. *Archives of environmental health,* **5**: 581–591 (1962).
3. COTE, I.L. & BAYARD, S.P. Cancer risk assessment of 1,3-butadiene. *Environmental health perspectives,* **86**: 149–153 (1990).
4. BELL, R.W. ET AL. *The 1990 Toronto Personal Exposure Pilot (PEP) Study.* Toronto, Ministry of the Environment, 1991.
5. LÖFROTH, G. ET AL. Characterization of environmental tobacco smoke. *Environmental science and technology,* **23**: 610–614 (1989).
6. BOND, J.A. ET AL. Epidemiological and mechanistic data suggest that 1,3-butadiene will not be carcinogenic to humans at exposures likely to be encountered in the environment or workplace. *Carcinogenesis,* **16**: 165–171 (1995).
7. MELNICK, R.L. & KOHN, M.C. Mechanistic data indicate that 1,3-butadiene is a human carcinogen. *Carcinogenesis,* **16**: 157–163 (1995).
8. SORSA, M. ET AL. Human cytogenetic biomonitoring of occupational exposure to 1,3-butadiene. *Mutation research,* **309**: 321–326 (1994).

9. US OCCUPATIONAL SAFETY AND HEALTH ADMINISTRATION. Occupational exposure to 1,3 butadiene: proposed rule and notice of hearing. *Federal register*, **55**: 32747 (1990).
10. BOND, J.A. ET AL. Species differences in the disposition of inhaled butadiene. *Toxicology and applied pharmacology*, **84**: 617–627 (1986).
11. *Proposed identification of 1,3-butadiene as a toxic air contaminant. Part B: health assessment.* Sacramento, California Air Resources Board, 1992.
12. *Integrated Risk Information System (IRIS)* (http://www.epa.gov/ngispgm3/iris/). Cincinnati, OH, US Environmental Protection Agency (accessed 18 September 2000).
13. SLOOFF, W. ET AL. *Exploratory report: 1,3-butadiene.* Bilthoven, National Institute of Public Health and Environmental Protection (RIVM), 1994 (Report No. 7104010333).
14. MELNICK, R.L. ET AL. Carcinogenicity of 1,3-butadiene in C57Bl/6 × C3HF1 mice at low exposure concentrations. *Cancer research*, **50**: 6592–6599 (1990).
15. OWEN, P.E. ET AL. Inhalation toxicity studies with 1,3-butadiene. 3. Two year toxicity/carcinogencity study in rats. *American Industrial Hygiene Association journal*, **48**: 407–413 (1987).
16. HIMMELSTEIN, M.W. ET AL. Comparison of blood concentrations of 1,3-butadiene and butadiene epoxides in mice and rats exposed to 1,3-butadiene by inhalation. *Carcinogenesis*, **15**: 1479–1486 (1994).
17. HIMMELSTEIN, M.W. ET AL. High concentrations of butadiene epoxides in livers and lungs of mice compared to rats exposed to 1,3-butadiene. *Toxicology and applied toxicology*, **132**: 281–288 (1995).
18. THORNTON-MANNING, J.R. ET AL. Disposition of butadiene monoepoxide and butadiene diepoxide in various tissues of rats and mice following a low-level inhalation exposure to 1,3-butadiene. *Carcinogenesis*, **16**: 1723–1731 (1995).

5.4 Carbon disulfide

Exposure evaluation

Inhalation represents the main route of entry of carbon disulfide into the human organism. Values in the vicinity of viscose rayon plants range from 0.01 mg/m^3 to about 1.5 mg/m^3, depending mostly on the distance from the source.

Health risk evaluation

A summary of the most relevant concentration–response findings is given in Table 10.

In the light of numerous epidemiological studies, it is very difficult to establish the exact exposure–time relationship. During the approximate period 1955–1965, carbon disulfide concentrations in viscose rayon plants averaged about 250 mg/m^3; they were subsequently reduced to 50–150 mg/m^3 and more recently to 20–30 mg/m^3. It is thus practically impossible to evaluate the long-term (five or more years) exposure level in a retrospective study. Moreover, most exposure data in occupational studies are not reliable, owing to poor measurement methodology. It is necessary to keep this in mind also when studying Table 10.

At exposure levels of 30 mg/m^3 and above, observable adverse health effects have been well established. The coronary heart disease rate increases at levels of 30–120 mg/m^3 of carbon disulfide after an exposure of more than 10 years. Effects on the central and peripheral nervous systems and the vascular system have been established in the same range of concentrations after long-term exposure. Functional changes of the central nervous system have even been observed at lower concentrations (20–25 mg/m^3).

Some authors claim to have observed adverse health effects in workers exposed to 10 mg/m^3 of carbon disulfide for 10–15 years. Because of the lack of reliable retrospective data on exposure levels, however, the dose–response relationship governing these findings is difficult to establish.

Guidelines

The lowest concentration of carbon disulfide at which an adverse effect was observed in occupational exposure was about 10 mg/m^3, which may be equivalent to a concentration in the general environment of 1 mg/m^3. In

Table 10. Some concentration–response relationships in occupational exposure to carbon disulfide

Carbon disulfide concentration (mg/m^3)	Duration of exposure (years)	Symptoms and signs	Reference
500–2500	0.5	Polyneuritis, myopathy, acute psychosis	*1*
450–1000	<0.5	Polyneuritis, encephalopathy	*2*
200–500	1–9	Increased ophthalmic pressure	*3*
60–175	5	Eye burning, abnormal pupillary light reactions	*4*
31–137	10	Psychomotor and psychological disturbances	*5*
29–118	15	Polyneuropathy, abnormal EEG, conduction velocity slowed, psychological changes	*6, 7*
29–118	10	Increase in coronary mortality, angina pectoris, slightly higher systolic and diastolic blood pressure	*8–11*
40–80	2	Asthenospermia, hypospermia, teratospermia	*12*
22–44	>10	Arteriosclerotic changes and hypertension	*13*
30–50	>10	Decreased immunological reactions	*14*
30	3	Increase in spontaneous abortions and premature births	*15*
20–25	<5	Functional disturbances of the central nervous system	*16, 17*
10	10–15	Sensory polyneuritis, increased pain threshold	*18*
10	10–15	Depressed blood progesterone, increased estriol, irregular menstruation	*19*

selecting the size of the protection (safety) factor, the expected variability in the susceptibility of the general population was taken into account, and a protection factor of 10 was considered appropriate. This leads to the recommendation of a guideline value of 100 μg/m³, with an averaging time of 24 hours. It is believed that below this value adverse health effects of environmental exposure to carbon disulfide (outdoor or indoor) are not likely to occur.

If carbon disulfide is used as the index substance for viscose emissions, odour perception is not to be expected when carbon disulfide peak concentration is kept below one tenth of its odour threshold value, i.e. below 20 μg/m³. Based on the sensory effects of carbon disulfide, a guideline value of 20 μg/m³ (average time 30 minutes) is recommended.

References

1. VIGLIANI, E.C. Chronic carbon disulfide poisoning: a report on 100 cases. *Medicina del lavoro,* **37**: 165–193 (1946).
2. VIGLIANI, E.C. Carbon disulphide poisoning in viscose rayon factories. *British journal of industrial medicine,* **11**: 235–244 (1954).
3. MAUGERI, U. ET AL. La oftalmodinamografia nella intossicazione solofocarbonica professionalle [Ophthalmodynamography in occupational carbon disulfide poisoning]. *Medicina del lavoro,* **57**: 730–740 (1966).
4. SAVIC, S. Influence of carbon disulfide on the eye. *Archives of environmental health,* **14**: 325–326 (1967).
5. HÄNNINEN, H. Psychological picture of manifest and latent carbon disulphide poisoning. *British journal of industrial medicine,* **28**: 374–381 (1971).
6. SEPPÄLAINEN, A.M. & TOLONEN, M. Neurotoxicity of long-term exposure to carbon disulfide in the viscose rayon industry: a neurophysiological study. *Work, environment, health,* **11**: 145–153 (1974).
7. TOLONEN, M. Chronic subclinical carbon disulfide poisoning. *Work, environment, health,* **11**: 154–161 (1974).
8. TOLONEN, M. ET AL. A follow-up study of coronary heart disease in viscose rayon workers exposed to carbon disulphide. *British journal of industrial medicine,* **32**: 1–10 (1975).
9. HERNBERG, S. ET AL. Excess mortality from coronary heart disease in viscose rayon workers exposed to carbon disulfide. *Work, environment, health,* **10**: 93–99 (1973).
10. NURMINEN, M. Survival experience of a cohort of carbon disulphide exposed workers from an eight-year prospective follow-up period. *International journal of epidemiology,* **5**: 179–185 (1976).

11. HERNBERG, S. ET AL. Coronary heart disease among workers exposed to carbon disulphide. *British journal of industrial medicine,* **27**: 313–325 (1970).
12. LANCRANJAN, I. ET AL. Changes of the gonadic function in chronic carbon disulfide poisoning. *Medicina del lavoro,* **60**: 556–571 (1969).
13. GAVRILESCU, N. & LILIS, R. Cardiovascular effects of long-extended carbon disulphide exposure. *In:* Brieger H. & Teisinger J., ed. *Toxicology of carbon disulphide.* Amsterdam, Excerpta Medica, 1967, pp. 165–167.
14. KAŠIN, L.M. [Overall immunological reactivity and morbidity of workers exposed to carbon disulfide]. *Gigiena i sanitarija,* **30**: 331–335 (1965) [Russian].
15. PETROV, M.V. [Course and termination of pregnancy in women working in the viscose industry]. *Pediatrija, akušerstvo i ginekologija,* **3**: 50–52 (1969) [Russian].
16. GILIOLI, R. ET AL. Study of neurological and neurophysiological impairment in carbon disulphide workers. *Medicina del lavoro,* **69**: 130–143 (1978).
17. CASSITO, M.G. ET AL. Subjective and objective behavioural alterations in carbon disulphide workers. *Medicina del lavoro,* **69**: 144–150 (1978).
18. MARTYNOVA, A.P. ET AL. [Clinical, hygienic and experimental investigations of the action on the body of small concentrations of carbon disulfide]. *Gigiena i sanitarija,* **5**: 25–28 (1976) [Russian].
19. VASILJEVA, I.A. [Effect of low concentrations of carbon disulfide and hydrogen sulfide on the menstrual function in women and on the estrous cycle under experimental conditions]. *Gigiena i sanitarija,* **38**: 24–27 (1973) [Russian].

5.5 Carbon monoxide

Exposure evaluation

Global background concentrations of carbon monoxide range between 0.06 mg/m^3 and 0.14 mg/m^3 (0.05–0.12 ppm). In urban traffic environments of large European cities, the 8-hour average carbon monoxide concentrations are generally lower than 20 mg/m^3 (17 ppm) with short-lasting peaks below 60 mg/m^3 (53 ppm). Carbon monoxide concentrations inside vehicles are generally higher than those measured in ambient outdoor air. The air quality data from fixed-site monitoring stations seem to reflect rather poorly short-term exposures of various urban population groups, but appear to reflect better longer averaging times, such as 8 hours.

In underground and multistorey car parks, road tunnels, enclosed ice arenas and various other indoor microenvironments, in which combustion engines are used under conditions of insufficient ventilation, the mean levels of carbon monoxide can rise above 115 mg/m^3 (100 ppm) for several hours, with short-lasting peak values that can be much higher. In homes with gas appliances, peak carbon monoxide concentrations of up to 60–115 mg/m^3 (53–100 ppm) have been measured. Environmental tobacco smoke in dwellings, offices, vehicles and restaurants can raise the 8-hour average carbon monoxide concentration to 23–46 mg/m^3 (20–40 ppm).

Carbon monoxide diffuses rapidly across alveolar, capillary and placental membranes. Approximately 80–90% of the absorbed carbon monoxide binds with haemoglobin to form carboxyhaemoglobin (COHb), which is a specific biomarker of exposure in blood. The affinity of haemoglobin for carbon monoxide is 200–250 times that for oxygen. During an exposure to a fixed concentration of carbon monoxide, the COHb concentration increases rapidly at the onset of exposure, starts to level off after 3 hours, and reaches a steady state after 6–8 hours of exposure. The elimination half-life in the fetus is much longer than in the pregnant mother.

In real-life situations, the prediction of individual COHb levels is difficult because of large spatial and temporal variations in both indoor and outdoor carbon monoxide concentrations.

Health risk evaluation

The binding of carbon monoxide with haemoglobin to form COHb reduces the oxygen-carrying capacity of the blood and impairs the release of oxygen from haemoglobin to extravascular tissues. These are the main causes of tissue hypoxia produced by carbon monoxide at low exposure levels. At higher concentrations the rest of the absorbed carbon monoxide binds with other haem proteins such as myoglobin, and with cytochrome oxidase and cytochrome P-450 *(1, 2)*. The toxic effects of carbon monoxide become evident in organs and tissues with high oxygen consumption such as the brain, the heart, exercising skeletal muscle and the developing fetus.

Severe hypoxia due to acute carbon monoxide poisoning may cause both reversible, short-lasting neurological deficits and severe, often delayed neurological damage. The neurobehavioural effects include impaired coordination, tracking, driving ability, vigilance and cognitive performance at COHb levels as low as 5.1–8.2% *(3–5)*.

In apparently healthy subjects, maximal exercise performance has decreased at COHb levels as low as 5%. The regression between the percentage decrease in maximal oxygen consumption and the percentage increase in COHb concentration appears to be linear, with a fall in oxygen consumption of approximately one percentage point for each percentage point rise in COHb level above 4% *(1, 6)*.

In controlled human studies involving patients with documented coronary artery disease, mean postexposure COHb levels of 2.9–5.9% (corresponding to postexercise COHb levels of 2.0–5.2%) have been associated with a significant shortening in the time to onset of angina, with increased electrocardiographic changes and with impaired left ventricular function during exercise *(7–11)*. In addition, ventricular arrhythmias may be increased significantly at the higher range of mean postexercise COHb levels *(12, 13)*. Epidemiological and clinical data indicate that carbon monoxide from recent smoking and environmental or occupational exposures may contribute to cardiovascular mortality and the early course of myocardial infarction *(1)*. According to one study there has been a 35% excess risk of death from arteriosclerotic heart disease among smoking and nonsmoking tunnel officers, in whom the long-term mean COHb levels were generally less than 5% *(13)*. Current data from epidemiological studies and experimental animal studies indicate that common environmental exposures to carbon monoxide do not have atherogenic effects on humans *(1, 14)*.

During pregnancy, endogenous production of carbon monoxide is increased so that maternal COHb levels are usually about 20% higher than the non-pregnant values. At steady state, fetal COHb levels are up to 10–15% higher than maternal COHb levels *(1, 15)*. There is a well established and probably causal relationship between maternal smoking and low birth weight at fetal COHb levels of 2–10%. In addition, maternal smoking seems to be associated with a dose-dependent increase in perinatal deaths and with behavioural effects in infants and young children *(15)*.

In contrast with most other man-made air pollutants at very high concentrations (well above ambient levels), carbon monoxide causes a large number of acute accidental and suicidal deaths in the general population.

Guidelines

In healthy subjects, endogenous production of carbon monoxide results in COHb levels of 0.4–0.7%. During pregnancy, elevated maternal COHb levels of 0.7–2.5%, mainly due to increased endogenous production, have been reported. The COHb levels in non-smoking general populations are usually 0.5–1.5%, owing to endogenous production and environmental exposures. Nonsmokers in certain occupations (car drivers, policemen, traffic wardens, garage and tunnel workers, firemen, etc.) can have long-term COHb levels of up to 5%, and heavy cigarette smokers have COHb levels of up to 10% *(1, 2, 15)*. Well trained subjects engaging in heavy exercise in polluted indoor environments can increase their COHb levels quickly up to 10–20%. In indoor ice arenas, epidemic carbon monoxide poisonings have recently been reported.

To protect nonsmoking, middle-aged and elderly population groups with documented or latent coronary artery disease from acute ischaemic heart attacks, and to protect the fetuses of nonsmoking pregnant women from untoward hypoxic effects, a COHb level of 2.5% should not be exceeded.

The following guidelines are based on the Coburn-Foster-Kane exponential equation, which takes into account all the known physiological variables affecting carbon monoxide uptake *(16)*. The following guideline values (ppm values rounded) and periods of time-weighted average exposures have been determined in such a way that the COHb level of 2.5% is not exceeded, even when a normal subject engages in light or moderate exercise:

- 100 mg/m^3 (90 ppm) for 15 minutes
- 60 mg/m^3 (50 ppm) for 30 minutes
- 30 mg/m^3 (25 ppm) for 1 hour
- 10 mg/m^3 (10 ppm) for 8 hours

References

1. *Air quality criteria for carbon monoxide*. Washington, DC, US Environmental Protection Agency, 1991 (Publication EPA-600/B-90/045F).
2. ACGIH Chemical Substances TLV Committee. Notice of intended change – carbon monoxide. *Applied occupational and environmental hygiene*, **6**: 621–624 (1991).
3. ACGIH Chemical Substances TLV Committee. Notice of intended change – carbon monoxide. *Applied occupational and environmental hygiene*, **6**: 896–902 (1991).
4. Putz, V.R. The effects of carbon monoxide on dual-task performance. *Human factors*, **21**: 13–24 (1979).
5. Benignus, V.A. et al. Effect of low level carbon monoxide on compensatory tracking and event monitoring. *Neurotoxicology and teratology*, **9**: 227–234 (1987).
6. Bascom, R. et al. Health effects of outdoor air pollution (Part 2). *American journal of respiratory and critical care medicine*, **153**: 477–498 (1996).
7. Anderson, E.W. et al. Effect of low-level carbon monoxide exposure on onset and duration of angina pectoris: a study in ten patients with ischemic heart disease. *Annals of internal medicine*, **79**: 46–50 (1973).
8. Kleinman, M.T. et al. Effects of short–term exposure to carbon monoxide in subjects with coronary artery disease. *Archives of environmental health*, **44**: 361–369 (1989).
9. Allred, E.N. et al. Short-term effects of carbon monoxide exposure on the exercise performance of subjects with coronary artery disease. *New England journal of medicine*, **321**: 1426–1432 (1989).
10. Sheps, D.S. et al. Lack of effect of low levels of carboxyhemoglobin on cardiovascular function in patients with ischemic heart disease. *Archives of environmental health*, **42**: 108–116 (1987).
11. Adams, K.F. et al. Acute elevation of blood carboxyhemoglobin to 6% impairs exercise performance and aggravates symptoms in patients with ischemic heart disease. *Journal of the American College of Cardiology*, **12**: 900–909 (1988).
12. Sheps, D.S. et al. Production of arrhythmias by elevated carboxyhemoglobin in patients with coronary artery disease. *Annals of internal medicine*, **113**: 343–351 (1990).
13. Stern, F.B. et al. Heart disease mortality among bridge and tunnel officers exposed to carbon monoxide. *American journal of epidemiology*, **128**: 1276–1288 (1988).
14. Smith, C.J. & Steichen, T.J. The atherogenic potential of carbon monoxide. *Atherosclerosis*, **99**: 137–149 (1993).

15. LONGO, L.D. The biological effects of carbon monoxide on the pregnant woman, fetus, and newborn infant. *American journal of obstetrics and gynecology*, **129**: 69–103 (1977).
16. COBURN, R.F. ET AL. Considerations of the physiological variables that determine the blood carboxyhemoglobin concentration in man. *Journal of clinical investigation*, **44**: 1899–1910 (1965).

5.6 1,2-Dichloroethane

Exposure evaluation

Rural or background atmospheric concentrations in western Europe and North America are approximately 0.2 μg/m^3, and the limited data available on indoor concentrations show that they are about the same. Average levels in cities vary from 0.4 μg/m^3 to 1.0 μg/m^3, increasing to 6.1 μg/m^3 near petrol stations, parking garages and production facilities.

Health risk evaluation

Human studies point to effects on the central nervous system and the liver, but the limited data do not allow a definitive conclusion regarding a LOAEL or NOAEL. In animals, long-term inhalation exposure (> 6 months) to 1,2-dichloroethane levels of approximately 700 mg/m^3 and above has been shown to result in histological changes in the liver *(1–3).* The same animal studies reported no adverse histological changes in the liver and kidneys of guinea pigs and rats at levels of about 400 mg/m^3. Findings concerning effects on reproduction are contradictory.

Animal data suggest a NOAEL in laboratory animals of 400 mg/m^3 and a LOAEL of 700 mg/m^3.

With regard to mutagenicity as an endpoint and to the causal connections between DNA damage and the initiation of carcinogenicity, 1,2-dichloroethane has been shown to be weakly mutagenic in *Salmonella typhimurium,* both in the absence and in the presence of microsomal activation systems. It has also been demonstrated to be mutagenic in other test species and in *in vitro* tests using mammalian cells.

In a lifetime study in rats and mice in which 1,2-dichloroethane was administered by gavage, it caused tumours at multiple sites in both species. In the only inhalation study performed *(4),* exposure to 1,2-dichloroethane did not result in an increased tumour incidence. The negative results obtained in this study, however, do not detract from the positive findings of the oral study *(5, 6)* when differences in total dose, exposure time and pharmacokinetics are considered.

1,2-Dichloroethane was evaluated in 1979 by IARC as a chemical for which there is sufficient evidence of carcinogenicity in experimental animals and inadequate evidence in humans *(7).* To date there are two publications giving quantitative carcinogenic risk estimates based on animal data. One, developed by the National Institute of Public Health in the Netherlands on the basis of oral exposure of rats by gavage *(6)*, indicates a lifetime risk of one in a million from exposure to 0.48 $\mu g/m^3$ *(8)*, which corresponds to a unit risk of about 2×10^{-6}. The US Environmental Protection Agency *(9)* has estimated an incremental unit risk of 2.6×10^{-5} on the basis of data from gavage studies and of 1×10^{-6} on the basis of a negative inhalation study.

Guidelines

Evidence of carcinogenicity in animals is sufficient on the basis of oral ingestion data. However, animal inhalation data do not at present provide positive evidence. Because of deficiencies in extrapolation from oral data to inhalation, the two risk estimates available are not used in the guidelines.

For noncarcinogenic endpoints, data from animal studies imply a NOAEL of about 400 mg/m^3 and suggest a LOAEL of about 700 mg/m^3. A protection (safety) factor of 1000 is considered appropriate in extrapolation of animal data to the general population. In selecting such a large protection factor, variations in exposure time, the limitations of the database and the fact that a no-effect level in humans cannot be established are of decisive importance. The resulting value of 0.7 mg/m^3 for continuous exposure (averaging time 24 hours) is recommended as a guideline value. Since this value is above current environmental levels and present exposures are not of concern to health, this guideline relates only to accidental release episodes or specific indoor pollution problems.

References

1. SPENCER, H.C. ET AL. Vapour toxicity of ethylene dichloride determined by experiments on laboratory animals. *Archives of industrial hygiene and occupational medicine*, 4: 482–493 (1951).
2. HEPPEL, L.A. ET AL. The toxicology of 1,2-dichloroethane (ethylene dichloride). V. The effect of daily inhalations. *Journal of industrial hygiene and toxicology*, 28: 113–120 (1946).
3. HOFMANN, H.T. ET AL. Zur Inhalationstoxizität von 1,1- und 1,2-Dichloräthan [On the inhalation toxicity of 1,1-and 1,2-dichloroethane]. *Archiv für Toxikologie*, 27: 248–265 (1971).
4. SPREAFICO, F. ET AL. Pharmacokinetics of ethylene dichloride in rats treated by different routes and its long-term inhalatory toxicity. *Banbury reports*, 5: 107–129 (1980).
5. US NATIONAL CANCER INSTITUTE. *Bioassay of 1,2-dichloroethane for possible carcinogenicity.* Bethesda, MD, US Department of Health, Education and Welfare, 1978 (DHEW Publication No. (NIH) 78-1305).

6. WARD, J.M. The carcinogenicity of ethylene dichloride in Osborne-Mendel rats and B6C3F1 mice. *Banbury reports,* 5: 35–53 (1980).
7. 1,2-Dichloroethane. *In: Some halogenated hydrocarbons.* Lyons, International Agency for Research on Cancer, 1979 (IARC Monographs on the Evaluation of the Carcinogenic Risk of Chemicals to Humans, Vol. 20), pp. 429–448.
8. BESEMER, A.C. ET AL. *Criteriadocument over 1,2-dichloorethaan* [1,2-Dichloroethane criteria document]. The Hague, Ministry of Housing, Spatial Planning and Environment, 1984 (Publikatiereeks Lucht, No. 30).
9. *Health assessment document for 1,2-dichloroethane (ethylene dichloride).* Washington, DC, US Environmental Protection Agency, 1985 (Report EPA-600/8-84-0067).

5.7 Dichloromethane

Exposure evaluation

Mean outdoor concentrations of dichloromethane are generally below 5 μg/m^3 *(1–4)*. Significantly higher concentrations (by at least one order of magnitude) may occur close to industrial emission sources. Indoor air concentrations are variable but tend to be about three times greater than outdoor values *(5, 6)*. Under certain circumstances, much higher values (up to 4000 μg/m^3) may be recorded indoors, particularly with use of paint stripping solutions *(7)*. Exposures of the general population occur principally through the use of dichloromethane-containing consumer products. Exposure in outdoor air, water *(8–12)* and food *(13–15)* is low.

Health risk evaluation

The critical effects of dichloromethane include effects on the central nervous system, the production of carboxyhaemoglobin (COHb) and carcinogenicity. The impairment of behavioural or sensory responses may occur in humans following acute inhalation exposure at levels exceeding 1050 mg/m^3 (300 ppm) for short durations, and the effects are transient. The cytochrome P-450-related oxidative pathway resulting in carbon monoxide production is saturable, producing maximum blood COHb levels of $\leq$ 9%. Nevertheless, these COHb levels are sufficiently high to induce acute effects on the central nervous system, and it thus appears that such effects are probably due to COHb production. Dichloromethane does not appear to cause serious effects in humans at those relatively high levels reported in occupational settings.

Although there is no convincing evidence of cancer incidence associated with occupational exposure, the available data have limitations and are considered inadequate to assess human carcinogenicity. In male and female mice and male and female rats, the National Toxicology Program's bioassays led to the conclusion of clear evidence of carcinogenicity in mice, clear evidence in female rats and equivocal evidence in male rats *(16)*. IARC has classified dichloromethane as showing sufficient evidence of carcinogenicity in experimental animals (Group 2B) *(17)*.

The health risks of exposure to dichloromethane have been considered in detail by an International Programme on Chemical Safety (IPCS) expert group. Given the data on interspecies differences in metabolism and comparative cancer risks, that group concluded that carcinogenicity was not the critical endpoint for risk assessment purposes. It is therefore concluded that the formation of COHb is a more direct indication of a toxic effect, that it can be monitored, and that it is therefore more suitable as a basis for the derivation of a guideline. Furthermore, it is unlikely that ambient air exposures represent a health concern with reference to any cancer endpoint, since concentrations of dichloromethane in ambient air are orders of magnitude lower than levels associated with direct adverse effects on the central nervous system or on COHb production in humans.

The application of physiologically based pharmacokinetic models to the available animal data lead to small risk estimates *(18, 19)*. These risk estimates are much lower than the recommended guideline value using COHb formation, and were therefore not employed in guideline derivation.

Guidelines

The selected biological endpoint of interest is the formation of COHb, which is measured in the blood of normal subjects at levels of 0.50–1.5% of total haemoglobin. In heavy smokers, the level of COHb may range up to 10%. Carbon monoxide from various sources may contribute to the formation of COHb. Since overall levels in many cases approach the recommended maximum of 3%, it is prudent to minimize any additional amounts of COHb contributed from dichloromethane. It was thus concluded that no more than 0.1% additional COHb should be formed from dichloromethane exposure. This corresponds to the analytical reproducibility of the method applied to measure COHb at the level of concern. This maximum allowable increase in COHb corresponds to a 24-hour exposure to dichloromethane at a concentration of 3 mg/m^3. Consequently, a guideline value of 3 mg/m^3 is recommended. In addition, the weekly average concentration should not exceed one seventh (0.45 mg/m^3) of this 24-hour guideline, given the half-life of COHb.

References

1. DANN, T. *Measurement of volatile organic compounds in Canada 1991–1992*. Ottawa, Environment Canada, 1993.
2. MCALLISTER, R. ET AL. *Non-methane organic compound program. Final Report, Vol. II*. Research Triangle Park, NC, US Environmental Protection Agency, 1989 (Report No. EPA-450/4-89-005).
3. SHIKIYA, J. ET AL. Ambient monitoring of selected halogenated hydrocarbons and benzene in the California South Coast Air Basin. *Proceedings of the Air Pollution Control Association 77th Annual Meeting*, Vol. 1, Paper 84e–1.1. Pittsburg, PA, Air Pollution Control Association, 1984.

4. Harkov, R. et al. Comparison of selected volatile organic compounds during the summer and winter at urban sites in New Jersey. *Science of the total environment,* **38**: 259–274 (1984).
5. Wallace, L. et al. The Los Angeles TEAM Study: personal exposures, indoor–outdoor air concentrations, and breath concentrations of 25 volatile organic compounds. *Journal of exposure analysis and environmental epidemiology*, **1**: 157–192 (1991).
6. Chan, C. C. et al. Determination of organic contaminants in residential indoor air using an adsorption-thermal desorption technique. *Journal of the Air and Waste Management Association*, **40**: 62–67 (1990).
7. Otson, R. et al. Dichloromethane levels in air after application of paint removers. *American Industrial Hygiene Association journal,* **42**: 56–60 (1981).
8. McGeorge, L. et al. Implementation and results of a mandatory statewide program for organic contaminant analysis of delivered water. *Proceedings of Water Quality Technology Conference,* **15**: 71–102 (1987).
9. *Priority Substances List Assessment Report. Dichloromethane*. Ottawa, Environment Canada and Health Canada, 1993.
10. Otson, R. et al. Volatile organic compounds in water at thirty Canadian potable water treatment facilities. *Journal of the Association of Official Analytical Chemists*, **65**: 1370–1374 (1982).
11. Otson, R. Purgeable organics in Great Lakes raw and treated water. *International journal of environmental analytical chemistry*, **31**: 41–53 (1987).
12. Staples, C.A. et al. Assessment of priority pollutant concentrations in the United States using STORET database. *Environmental toxicology and chemistry*, **4**: 131–142 (1985).
13. Ferrario, J.B. et al. Volatile organic pollutants in biota and sediments of Lake Pontchartrain. *Bulletin of environmental contamination and toxicology*, **43**: 246–255 (1985).
14. Heikes, D.L. Environmental contaminants in table-ready foods from the total diet program of the Food and Drug Administration. *Advances in environmental science and technology,* **23**: 31–57 (1990).
15. Page, B.D. & Charbonneau, C.F. Headspace gas chromatographic determination of residual methylene chloride in decaffeinated tea and coffee with electronic conductivity detection. *Journal of the Association of Official Analytical Chemists*, **67**: 757–761 (1984).
16. National Toxicology Program. *Toxicology and carcinogenesis studies of dichloromethane (methylene chloride) (CAS No. 75-09-2) in F344/N rats and B6C3F1 mice (inhalation studies).* Research Triangle Park, NC, US Department of Health and Human Services, 1986 (Document No. NTP-TRS-306).

17. *Some halogenated hydrocarbons and pesticide exposure*. Lyons, International Agency for Research on Cancer, 1986 (IARC Monographs on the Evaluation of the Carcinogenic Risk of Chemicals to Humans, Vol. 41), pp. 43–85.
18. *Methylene chloride (dichloromethane): human risk assessment using experimental animal data*. Brussels, European Centre for Ecotoxicology and Toxicology of Chemicals, 1988 (Technical Report No.32).
19. DANKOVIC, D.A. & BAILER, A.J. The impact of exercise and intersubject variability on dose estimates for dichloromethane derived from a physiologically based pharmacokinetic model. *Fundamental and applied toxicology*, **22**: 20–25 (1994).

5.8 Formaldehyde

Exposure evaluation

The major route of exposure to formaldehyde is inhalation. Table 11 shows the contribution of the various atmospheric environments to non-occupational air levels. Indoor air concentrations are several orders of magnitude higher than levels in ambient air. Owing to the extremely high concentrations of formaldehyde in tobacco smoke, smoking constitutes a major source of formaldehyde *(1)*.

Health risk evaluation

Predominant symptoms of formaldehyde exposure in humans are irritation of the eyes, nose and throat, together with concentration-dependent discomfort, lachrymation, sneezing, coughing, nausea, dyspnoea and finally death (Table 12) .

Damage to the nasal mucosa, such as squamous cell metaplasia and mild dysplasia of the respiratory epithelium, have been reported in humans, but

Table 11. Average exposure concentrations to formaldehyde and contribution of various atmospheric environments to average exposure to formaldehyde

Source	Concentration (mg/m³)	Exposure (mg/day)
Ambient air (10% of time; 2 m³/day)	0.001–0.02	0.002–0.04
Indoor air		
Home (65% of time; 10 m³/day)		
– conventional	0.03–0.06	0.3–0.6
– mobile home	0.1	1.0
– environmental tobacco smoke	0.05–0.35	0.5–3.5
Workplace (25% of time; 8 m³/day)		
– without occupational exposure [a]	0.03–0.06	0.2–0.5
– with occupational exposure	1.0	8.0
– environmental tobacco smoke	0.05–0.35	0.4–2.8
Smoking (20 cigarettes/day)	60–130	0.9–2.0 [b]

[a] Assuming the normal formaldehyde concentration in conventional buildings.

[b] Total amount of formaldehyde in smoke from 20 cigarettes.

Source: World Health Organization *(2)*.

Table 12. Effects of formaldehyde in humans after short-term exposure

Concentration range or average (mg/m^3)	Time range or average	Health effects in general population
0.03	Repeated exposure	Odour detection threshold (10th percentile) [a]
0.18	Repeated exposure	Odour detection threshold (50th percentile) [a]
0.6	Repeated exposure	Odour detection threshold (90th percentile) [a]
0.1–3.1	Single and repeated exposure	Throat and nose irritation threshold
0.6–1.2	Single and repeated exposure	Eye irritation threshold
0.5–2.0	3–5 hours	Decreased nasal mucus flow rate
2.4	40 minutes on 2 successive days with 10 minutes of moderate exercise on second day	Postexposure (up to 24 hours) headache
2.5–3.7	–[b]	Biting sensation in eyes and nose
3.7	Single and repeated exposure	Decreased pulmonary function only at heavy exercise
5–6.2	30 minutes	Tolerable for 30 minutes with lachrymation
12–25	–[b]	Strong lachrymation, lasting for 1 hour
37–60	–[b]	Pulmonary oedema, pneumonia, danger to life
60–125	–[b]	Death

[a] Frequency of effect in population.

[b] Time range or average unspecified.

these findings may have been confounded by concomitant exposures to other substances *(3)*.

There is convincing evidence of high concentrations of formaldehyde being capable of inducing nasal cancer in rats and possibly in mice *(3)*. Formaldehyde has been shown to be genotoxic in a variety of *in vitro* and *in vivo* systems *(3)*. There is also epidemiological evidence of associations between relatively high occupational exposure to formaldehyde and both nasopharyngeal and sinonasal cancers *(3–7)*.

There is substantial variation in individual responses to formaldehyde in humans *(1–3)*. Significant increases in signs of irritation occur at levels above 0.1 mg/m^3 in healthy subjects. At concentrations above 1.2 mg/m^3, a progression of symptoms and effects occurs. Lung function of healthy nonsmokers and asthmatics exposed to formaldehyde at levels up to 3.7 mg/m^3 was generally unaltered *(8–10)*. It is assumed that in these studies the observed effects were more related to peak concentrations than to mean values.

There is some evidence of formaldehyde inducing pathological and cytogenetic changes in the nasal mucosa of humans. Reported mean exposures ranged from 0.02 mg/m^3 to 2.4 mg/m^3, with peaks between 5 mg/m^3 and 18 mg/m^3. Epidemiological studies suggest a causal relationship between exposure to formaldehyde and nasopharyngeal cancer, although the conclusion is tempered by the small numbers of observed and expected cases *(3–6)*. There are also epidemiological observations of an association between relatively high occupational exposures to formaldehyde and sinonasal cancer *(7)*. IARC *(3)* has interpreted the available cancer data as limited evidence for the carcinogenicity of formaldehyde in humans, and classified formaldehyde in Group 2A.

Formaldehyde is a nasal carcinogen in rats. A highly significant incidence of nasal cancer was found in rats exposed to a level of 16.7 mg/m^3, but the dose–response curve was nonlinear, the risk being disproportionately low at low concentrations. It also appears that the dose–response curves were nearly identical for neoplastic changes, cell turnover, DNA–protein crosslinks and hyperproliferation, when the relationship between non-neoplastic and neoplastic lesions in the nasal respiratory epithelium was analysed. This close concordance indicates an association among the observed cytotoxic, genotoxic and carcinogenic effects. It is thus likely that hyperproliferation induced by cytotoxicity plays a significant role in the formation of nasal tumours by formaldehyde.

Despite differences in the anatomy and physiology of the respiratory tract between rats and humans, the respiratory tract defence mechanisms are similar. It is therefore reasonable to assume that the response of the human respiratory tract mucosa to formaldehyde will be similar to that of the rat. Thus, if the respiratory tract tissue is not repeatedly damaged, exposure of humans to low, noncytotoxic concentrations of formaldehyde can be assumed to be associated with a negligible cancer risk. This is consistent with epidemiological findings of excess risks of nasopharyngeal and sinonasal cancers associated with concentrations above about 1 mg/m^3.

Simultaneous exposure of humans to formaldehyde and other upper respiratory tract toxicants, such as acrolein, acetaldehyde, crotonaldehyde, furfural, glutaraldehyde and ozone, may lead to additive or synergistic effects, in particular with respect to sensory irritation and possibly also regarding cytotoxic effects on the nasal mucosa *(3, 11–16)*.

Guidelines

The lowest concentration that has been associated with nose and throat irritation in humans after short-term exposure is 0.1 mg/m^3, although some individuals can sense the presence of formaldehyde at lower concentrations.

To prevent significant sensory irritation in the general population, an air quality guideline value of 0.1 mg/m^3 as a 30-minute average is recommended. Since this is over one order of magnitude lower than a presumed threshold for cytotoxic damage to the nasal mucosa, this guideline value represents an exposure level at which there is a negligible risk of upper respiratory tract cancer in humans.

References

1. *Air quality guidelines for Europe*. Copenhagen, WHO Regional Office for Europe, 1987 (WHO Regional Publications, European Series, No. 23).
2. *Formaldehyde.* Geneva, World Health Organization, 1989 (Environmental Health Criteria, No. 89).
3. Formaldehyde. *In*: *Wood dust and formaldehyde.* Lyons, International Agency for Research on Cancer, 1995 (IARC Monographs on the Evaluation of Carcinogenic Risks to Humans, Vol. 62), pp. 217–362.
4. Blair, A. et al. Epidemiologic evidence on the relationship between formaldehyde exposure and cancer. *Scandinavian journal of work, environment and health*, **16**: 381–393 (1990).
5. Partanen, T. Formaldehyde exposure and respiratory cancer – a meta-analysis of the epidemiologic evidence. *Scandinavian journal of work, environment and health*, **19**: 8–15 (1993).
6. McLaughlin, J.K. Formaldehyde and cancer: a critical review. *International archives of occupational and environmental health*, **66**: 295–301 (1994).

7. HANSEN, J. & OLSEN, J.H. Formaldehyde and cancer morbidity among male employees in Denmark. *Cancer causes and control*, **6**: 354–360 (1995).
8. SAUNDER, L.R. ET AL. Acute pulmonary response to formaldehyde exposure in healthy nonsmokers. *Journal of occupational medicine*, **28**: 420–424 (1986).
9. SAUNDER, L.R. ET AL. Acute pulmonary response of asthmatics to 3.0 ppm formaldehyde. *Toxicology and industrial health*, **3**: 569–578 (1987).
10. GREEN, D.J. ET AL. Acute response to 3.0 ppm formaldehyde in exercising healthy nonsmokers and asthmatics. *American review of respiratory diseases*, **135**: 1261–1266 (1987).
11. CASSEE, F.R. & FERON, V.J. Biochemical and histopathological changes in nasal epithelium of rats after 3-day intermittent exposure to formaldehyde and ozone alone or in combination. *Toxicology letters*, **72**: 257–268 (1994).
12. LAM, C.-W. ET AL. Depletion of nasal mucosal glutathione by acrolein and enhancement of formaldehyde-induced DNA–protein cross-linking by simultaneous exposure to acrolein. *Archives of toxicology*, **58**: 67–71 (1985).
13. CHANG, J.C.F. & BARROW C.S. Sensory irritation tolerance and cross-tolerance in F-344 rats exposed to chlorine or formaldehyde gas. *Toxicology and applied pharmacology*, **76**: 319–327 (1984).
14. BABIUK, C. ET AL. Sensory irritation response to inhaled aldehydes after formaldehyde pretreatment. *Toxicology and applied pharmacology*, **79**: 143–149 (1985).
15. GRAFSTRÖM, R.C. ET AL. Genotoxicity of formaldehyde in cultured human bronchial fibroblasts. *Science*, **228**: 89–91 (1985).
16. GRAFSTRÖM, R.C. ET AL. Mutagenicity of formaldehyde in Chinese hamster lung fibroblasts: synergy with ionizing radiation and N-nitroso-N-methylurea. *Chemical–biological interactions*, **86**: 41–49 (1993).

5.9 Polycyclic aromatic hydrocarbons

Exposure evaluation

Polycyclic aromatic hydrocarbons (PAHs) are formed during incomplete combustion or pyrolysis of organic material and in connection with the worldwide use of oil, gas, coal and wood in energy production. Additional contributions to ambient air levels arise from tobacco smoking, while the use of unvented heating sources can increase PAH concentrations in indoor air. Because of such widespread sources, PAHs are present almost everywhere. PAHs are complex mixtures of hundreds of chemicals, including derivatives of PAHs, such as nitro-PAHs and oxygenated products, and also heterocyclic PAHs. The biological properties of the majority of these compounds are as yet unknown. Benzo[*a*]pyrene (BaP) is the PAH most widely studied, and the abundance of information on toxicity and occurrence of PAHs is related to this compound. Current annual mean concentrations of BaP in major European urban areas are in the range 1–10 ng/m^3. In rural areas, the concentrations are < 1 ng/m^3 *(1–5)*.

Food is considered to be the major source of human PAH exposure, owing to PAH formation during cooking or from atmospheric deposition of PAHs on grains, fruits and vegetables. The relative contribution of airborne PAH pollutants to food levels (via fallout) has not been well characterized *(6)*.

Health risk evaluation

Data from animal studies indicate that several PAHs may induce a number of adverse effects, such as immunotoxicity, genotoxicity, carcinogenicity and reproductive toxicity (affecting both male and female offspring), and may possibly also influence the development of atherosclerosis. The critical endpoint for health risk evaluation is the well documented carcinogenicity of several PAHs *(7)*.

BaP is by far the most intensively studied PAH in experimental animals. It produces tumours of many different tissues, depending on the species tested and the route of application. BaP is the only PAH that has been tested for carcinogenicity following inhalation, and it produced lung tumours in hamsters, the only species tested. Induction of lung tumours in rats and hamsters has also been documented for BaP and several other PAHs following direct application, such as intratracheal instillation into the pulmonary tissue. The lung carcinogenicity of BaP can be enhanced by coexposure to other substances such as cigarette smoke, asbestos and probably also airborne particles. Several studies have shown that the benzene-soluble fraction, containing 4- to 7-ring PAHs of condensates from car exhausts, domestic coal-stove emissions and tobacco smoke, contains nearly all the carcinogenic potential of PAHs from these sources *(8)*.

Because several PAHs have been shown to be carcinogenic, and many more have been shown to be genotoxic in *in vitro* assays, a suitable indicator for the carcinogenic fraction of the large number of PAHs in ambient air is desirable. The most appropriate indicator for the carcinogenic PAHs in air seems to be BaP concentrations, given present knowledge and the existing database. Assessment of risks to health of a given mixture of PAHs using this indicator approach would entail, first, measurement of the concentration of BaP in a given mixture present in a medium such as air. Then, assuming that the given mixture resembles that from coke ovens, the unit risk estimate is applied in tandem with the measured BaP air concentration to obtain the lifetime cancer risk at this exposure level.

The proportions of different PAHs detected in different emissions and workplaces sometimes differ widely from each other and from PAH profiles in ambient air. Nevertheless, the profiles of PAHs in ambient air do not seem to differ very much from one area to another, although large variations may be seen under special conditions. Furthermore, the carcinogenicity of PAH mixtures may be influenced by synergistic and antagonistic effects of other compounds emitted together with PAHs during incomplete combustion. It should also be recognized that in ambient air the carcinogenic 4- to 7-ring PAHs (representing the majority of PAHs) are preferentially attached to particles and only a minor fraction, depending on the temperature, exists as volatiles. A few studies indicate that the toxicokinetic properties of inhaled BaP attached to particles are different from those of pure BaP alone. Virtually nothing is known about other PAHs in this respect.

Risk assessments and potency assessments of various individual PAHs and complex mixtures of PAHs have been attempted. BaP is the only PAH for which a database is available, allowing a quantitative risk assessment. Risk assessment of BaP is, however, hampered by the poor quality of the data sets available *(9)*.

Attempts to derive relative potencies of individual PAHs (relative to BaP) have also been published, and the idea of summarizing the contributions from each of the selected PAHs into a total BaP equivalent dose (assuming their carcinogenic effects to be additive) has emerged *(10, 11)*. There are doubts, however, about the scientific justification for these procedures.

Risk estimates considered in the United States for coke-oven emissions were used in the first edition of these guidelines. Using a linearized multistage model, the most plausible upper-bound individual lifetime unit risk estimate associated with a continuous exposure to 1 $\mu g/m^3$ of benzene-soluble compounds of coke-oven emissions in ambient air was approximately 6.2×10^{-4}. Using BaP as an indicator of general PAH mixtures from emissions of coke ovens and similar combustion processes in urban air, and a reported value of 0.71% BaP in the benzene-soluble fraction of coke oven emissions, a lifetime risk of respiratory cancer of 8.7×10^{-5} per ng/m^3 was calculated *(1)*.

From the lung tumour rates obtained in a recent rat inhalation study with coal tar/pitch condensation aerosols, containing two different levels of BaP, a lifetime tumour risk of 2×10^{-5} per ng/m^3 for BaP as a constituent of a complex mixture was calculated using a linearized multistage model *(12)*.

Guidelines

No specific guideline value can be recommended for PAHs as such in air. These compounds are typically constituents of complex mixtures. Some PAHs are also potent carcinogens, which may interact with a number of other compounds. In addition, PAHs in air are attached to particles, which may also play a role in their carcinogenicity. Although food is thought to be the major source of human exposure to PAHs, part of this contamination may arise from air pollution with PAHs. The levels of PAHs in air should therefore be kept as low as possible.

In view of the difficulties in dealing with guidelines for PAH mixtures, the advantages and disadvantages of using a single indicator carcinogen to represent the carcinogenic potential of a fraction of PAH in air were considered. Evaluation of, for example, BaP alone will probably underestimate the carcinogenic potential of airborne PAH mixtures, since co-occurring substances are also carcinogenic. Nevertheless, the well studied common constituent of PAH mixtures, BaP, was chosen as an indicator, although the limitations and uncertainties in such an approach were recognized.

To set priorities with respect to control, an excess lifetime cancer risk, expressed in terms of the BaP concentration and based on observations in coke-oven workers exposed to mixtures of PAHs, is presented here. It must be emphasized that the composition of PAHs to which coke-oven workers are exposed may not be similar to that in ambient air, although it was noted that similar risks have been derived from studies of individuals exposed to other mixtures containing PAHs. Having also taken into consideration some recent animal data from which a unit risk of the same order of magnitude can be derived, it was concluded that the occupational epidemiology data should serve as the basis for the risk estimate.

Based on epidemiological data from studies in coke-oven workers, a unit risk for BaP as indicator air constituent for PAHs is estimated to be 8.7×10^{-5} per ng/m^3, which is the same as that established by WHO in 1987. The corresponding concentrations of BaP producing excess lifetime cancer risks of 1/10 000, 1/100 000 and 1/1 000 000 are 1.2, 0.12 and 0.012 ng/m^3, respectively.

References

1. Polynuclear aromatic hydrocarbons (PAH). *In*: *Air quality guidelines for Europe*. Copenhagen, WHO Regional Office for Europe, 1987, pp. 105–117.
2. *Toxicological profile for polycyclic aromatic hydrocarbons (PAHs): update.* Atlanta, GA, Agency for Toxic Substances and Disease Registry, 1994.
3. BAEK, S.O. ET AL. A review of atmospheric polycyclic aromatic hydrocarbons: sources, fate and behavior. *Water, air, and soil pollution,* **60**: 279–300 (1991).
4. PFEFFER, H.U. Ambient air concentrations of pollutants at traffic-related sites in urban areas of North Rhine-Westphalia, Germany. *Science and the total environment,* **146/147**: 263–273 (1994).
5. NIELSEN, T. ET AL. *Traffic PAH and other mutagens in air in Denmark.* Copenhagen, Danish Environmental Protection Agency, 1995 (Miljøprojekt No. 285).
6. DE VOS, R.H. ET AL. Polycyclic aromatic hydrocarbons in Dutch total diet samples (1984–1986). *Food chemistry and toxicology,* **28**: 263–268 (1990).
7. *Polynuclear aromatic compounds. Part 1. Chemical, environmental and experimental data.* Lyons, International Agency for Research on Cancer, 1983 (IARC Monographs on the Evaluation of the Carcinogenic Risk of Chemicals to Humans, Vol. 32).
8. POTT, F. & HEINRICH, U. Relative significance of different hydrocarbons for the carcinogenic potency of emissions from various incomplete combustion processes. *In*: Vainio, H. et al., ed. *Complex mixtures and cancer risk.* Lyons, International Agency for Research on Cancer, 1990, pp. 288–297 (IARC Scientific Publications, No. 104).
9. COLLINS, J.F. ET AL. Risk assessment for benzo[*a*]pyrene. *Regulatory toxicology and pharmacology,* **13**: 170–184 (1991).
10. NISBET, I.C.T. & LAGOY, K. Toxic equivalency factors (TEFs) for polycyclic aromatic hydrocarbons (PAHs). *Regulatory toxicology and pharmacology,* **16**: 290–300 (1992).

11. RUGEN, P.J. ET AL. Comparative carcinogenicity of the PAHs as a basis for acceptable exposure levels (AELs) in drinking water. *Regulatory toxicology and pharmacology*, **9**: 273–283 (1989).
12. HEINRICH, U. ET AL. Estimation of a lifetime unit lung cancer risk for benzo[*a*]pyrene based on tumour rates in rats exposed to coal tar/pitch condensation aerosol. *Toxicology letters*, **72**: 155–161 (1994).

5.10 Polychlorinated biphenyls

Exposure evaluation

Analysis of polychlorinated biphenyls (PCBs) should be performed by congener-specific methods. The method of quantifying total PCBs, by comparing the sample peak pattern with that of a commercial mixture, is accurate only when the sample under investigation has been directly contaminated by a commercial mixture. Because of substantial differences in PCB patterns between biological samples and technical products, however, this method leads to errors in the quantification of biological samples and also to differences between laboratories owing to the use of different standard mixtures. As a consequence, data have to be interpreted with great care. Comparisons can only be made between data either from the same laboratory, using the same validated technique and the same standards over a longer period, or from different laboratories when very strict interlaboratory controls have been applied. Indications of trends can only be obtained when these considerations are taken into account.

Food

Food is the main source of human intake of PCBs; intake through drinking-water is negligible.

The daily intake of total PCBs in Sweden was recently estimated at 0.05 µg/kg body weight (BW), with a 50% contribution from fish *(1)*. This is markedly lower than an earlier Finnish estimate of 0.24 µg/kg BW *(2)*, and might reflect the decreasing trends in PCB levels in Nordic food. Recent data from the Nordic countries indicate that the current average daily intake in toxic equivalents of dioxin-like PCBs may be slightly above 1 pg/kg BW *(3, 4)*.

If the contributions of PCDDs and PCDFs are also taken into account, the daily intake in toxic equivalents would be in the range 2–6 pg/kg BW for many European countries and the United States *(5)*. For certain risk groups, such as fishermen from the Baltic Sea and Inuits in the Arctic who consume large amounts of contaminated fatty fish, the intake may be up to four times higher.

Air

PCB levels have been shown to be higher in indoor air than in ambient air. Inhalation exposure to PCBs, assuming an indoor air level of 3 ng/m^3 in an uncontaminated building and an inhaled volume of 20 m^3 of air per day for adults, is approximately 0.001 μg/kg BW per day. In contaminated buildings concentrations above 300 ng/m^3 have been found, corresponding to a daily dose of at least 0.1 μg/kg BW. In buildings using PCB-containing sealants, levels up to 7500 ng/m^3 have been found (corresponding to a daily dose of 2.5 μg/kg BW). In ambient air there is a wide variation in the measurements from nonindustrialized (e.g. 0.003 ng/m^3) and industrial/urban areas (e.g. 3 ng/m^3). The levels of dioxin-like PCBs cannot be estimated owing to the lack of congener-specific analytical data.

Health risk evaluation

In 1990, the Joint FAO/WHO Expert Committee on Food Additives concluded that, owing to the limitations of the available data, it was impossible to establish a precise numerical value for a tolerable intake of total PCBs for humans *(6)*. IARC concluded that available studies suggested an association between human cancer and exposure to PCBs *(7)*. Overall, PCBs were classified as probably carcinogenic to humans (Group 2A), although several national governments are employing tolerable daily intakes (TDIs) for PCBs for the purpose of risk management.

In Germany a TDI for PCB of 1–3 μg/kg BW has been suggested. It was also recommended that, for precautionary reasons, the proportional daily intake via indoor air should not exceed 10% of the TDI for long periods. On this basis, an action level for source removal of 3000 μg/m^3 has been derived. For concentrations between 30 000 μg/m^3 and 10 000 μg/m^3, a concrete health risk is not assumed. However, mitigation measures should be undertaken as soon as possible to reduce the level to 300 μg/m^3, below which concentrations are thought to be of no concern. Source removal should also be undertaken if levels are found to be between 300 and 3000 μg/m^3 *(8)*.

Neurobehavioural and hormonal effects have been observed in infants exposed to background concentrations of PCBs, prenatally and/or through breastfeeding. The clinical significance of these observations is, however, unclear.

On average, the contribution from inhalation exposure is approximately 1% of the dietary intake but may approach that intake in certain extreme situations (areas close to sources or contaminated indoor air).

Exposures to dioxin-like PCBs can be converted to toxic equivalents using the WHO/IPCS interim toxic equivalent factors *(9)* and subsequently be

Air quality guidelines for Europe, 2nd edition, 2000

CORRIGENDUM

On page 98, under Health risk evaluation, the second paragraph should read as follows.

In Germany a TDI for PCB of 1–3 μg/kg BW has been suggested. It was also recommended that, for precautionary reasons, the proportional daily intake via indoor air should not exceed 10% of the TDI for long periods. On this basis an action level for source removal of 3000 ng/m^3 has been derived. For concentrations between 3000 ng/m^3 and 10 000 ng/m^3 (that is, between 3 $\mu g/m^3$ and 10 $\mu g/m^3$) a concrete health risk is not assumed. However, mitigation measures should be undertaken as soon as possible to reduce the level to 300 ng/m^3, below which concentrations are thought to be of no concern. Source removal should also be undertaken if levels are found to be between 300 and 3000 ng/m^3 *(8)*.

assessed using the TDI for TCDD. In 1992, WHO established a TDI for TCDD of 10 pg/kg BW. This was derived on the basis of TCDD-induced liver cancer in rats *(10)* for which a NOAEL of 1 ng/kg BW per day, corresponding to a liver concentration of 540 ng/kg on a wet-weight basis, was calculated. Owing to toxicokinetic differences between humans and rats, this would correspond to a daily intake in humans of 100 pg/kg BW, to which value an uncertainty factor of 10 (to cover inter-individual variation) was applied. Although not explicitly stated, the TDI can be looked on as applicable to the total intake of toxic equivalents derived from PCDDs, PCDFs and other dioxin-like compounds that act by the same mechanisms and cause similar types of toxicity.

For the average consumer, the daily intake of dioxin-like PCBs determined as toxic equivalents would be 10–30% of the TDI. When the contribution from the PCDDs and PCDFs is taken into account, the intake would increase to 20–60%. There are, however, groups with specific dietary habits (such as a high intake of contaminated food) or occupational exposure that may exceed the TDI for PCDDs and PCDFs.

The WHO human milk exposure study *(11)* indicated that the daily intake in toxic equivalents of PCDDs and PCDFs in breastfeeding infants in industrialized countries ranged from about 20 pg/kg BW in less industrialized areas to about 130 pg/kg BW in highly industrialized areas. This indicates intakes 2–13 times higher than the TDI. When the contribution from dioxin-like PCBs is taken into account, the intakes may be up to 2 times higher. It has been noted, however *(12)*, that the TDI should not be applied to such infants because the TDI concept relates to a dose ingested throughout a lifetime. The quantity of PCDDs and PCDFs ingested over a 6-month breastfeeding period would be less than 5% of the quantity ingested over a lifetime.

Guidelines

An air quality guideline for PCBs is not proposed because direct inhalation exposures constitute only a small proportion of the total exposure, in the order of 1–2% of the daily intake from food. WHO has not developed a TDI for total PCB exposure. Owing to the multiplicity of mechanisms underlying PCB-induced health effects, there may not be a scientifically sound rationale to set such a TDI. Average ambient air concentrations of PCBs are estimated to be 3 ng/m^3 in urban areas. Although this air concentration is only a minor contributor to direct human exposure, it is a major contributor to contamination of the food chain. It would also be possible to perform such calculations using toxic equivalents for dioxin-like PCBs in ambient air, but no such analytical data have been published.

Although indoor air levels of PCBs are generally very low, in certain instances levels of up to several μg/m^3 have been detected. For people living or working in such buildings, exposure to PCBs via air could contribute significantly to the overall PCB exposure.

Because of the potential importance of the indirect contribution of PCBs in air to total human exposure, it is important to control known sources as well as to identify new sources.

References

1. DARNERUD, P.O. ET AL. Bakgrund till de reviderade kostråden. PCB och dioxiner i fisk [Background to the revised dietary guidelines. PCB and dioxins in fish]. *Vår föda*, **47**: 10–21 (1995).
2. MOILANEN, R. ET AL. Average total dietary intakes of organochlorine compounds from the Finnish diet. *Zeitung für Lebensmittel Untersuchung und Forschung*, **182**: 484–488 (1986).
3. FÆRDEN, K. *Dioksiner i næringsmidler*. [Dioxins in food]. Oslo, National Food Control Authority, 1991 (SNT-Rapport No. 4).
4. SVENSON, B.G. ET AL. Exposure to dioxins and dibenzofurans through consumption of fish. *New England journal of medicine*, 324; 8–1 (1991).
5. *International toxicity equivalency factors (I–TEF) method of risk assessment for complex mixtures of dioxins and related compounds*. Brussels, North Atlantic Treaty Organization, 1988 (Report No. 176).
6. *Evaluation of certain food additives and contaminants. Thirty-fifth report of the Joint FAO/WHO Expert Committee on Food Additives*. Geneva, World Health Organization, 1990 (WHO Technical Report Series, No. 789).
7. *Overall evaluations of carcinogenicity: an updating of IARC monographs volumes 1 to 42*. Lyons, International Agency for Research on Cancer, 1987 (IARC Monographs on the Evaluation of the Carcinogenic Risk of Chemicals to Humans, Supplement 7).
8. ROSSKAMP, E. Polychloriente Biphenyle in der Innerraumluft – Sachstand. *Bundesgesundheitsblatt*, **35**: 434 (1992).
9. AHLBORG, U.G. ET AL. Toxic equivalency factors for dioxin-like PCBs. Report on a WHO-ECEH and IPCS consultation, December 1993. *Chemosphere*, **28**: 1049–1067 (1994).
10. KOCIBA, R.J. ET AL. Results of a two year chronic toxicity and oncogenicity study of 2,3,7,8-tetrachlordibenzo-*p*-dioxin (TCDD) in rats. *Toxicology and applied pharmacology*, **46**: 279–303 (1978).

11. *Levels of PCBs, PCDDs and PCDFs in human milk. Second round of WHO coordinated exposure study.* Geneva, World Health Organization, 1992 (Environmental Health Series, No.3).
12. AHLBORG, U.G. ET AL., ED. Special issue: tolerable daily intake of PCDDs and PCDFs. *Toxic substances journal,* **12**: 101–131 (1992).

5.11 Polychlorinated dibenzodioxins and dibenzofurans

Exposure evaluation

Food is the main source of human intake of polychlorinated dibenzodioxins (PCDDs) and dibenzofurans (PCDFs); intake through drinking-water is negligible. Calculated as toxic equivalents, average intakes in European countries have been estimated to be in the range 1.5–2 pg/kg body weight (BW) per day *(1–3)*. Very recent data from the Nordic countries indicate that this figure today may be slightly less than 1 pg/kg BW per day *(4, 5)*. For the United States, intake estimates are in the range 1–3 pg/kg BW per day *(6)*.

If the contributions of dioxin-like polychlorinated biphenyls (PCBs) are taken into account, and using the WHO toxic equivalency factors (TEFs) for PCBs *(7, 8)*, the toxic equivalent intake would be in the range 2–6 pg/kg BW per day. For certain risk groups, such as fishermen from the Baltic Sea and Inuits in the Arctic, intakes may be considerably higher.

Inhalation exposure to PCDDs and PCDFs is generally low. Assuming an ambient air toxic equivalent level of 0.1 pg/m^3 and an inhaled volume of air of 20 m^3/day for adults, inhalation intake would amount to about 0.03 pg/kg BW per day *(9, 10)*. Certain industrial and urban areas, however, as well as areas close to major sources, may have up to 20 times higher air concentrations. The contribution to the total toxic equivalents of dioxin-like PCBs from ambient air cannot be calculated owing to lack of congener-specific data. Under special circumstances, for example indoor air highly contaminated from coated particle boards containing PCBs, inhalation exposure may reach 1 pg/kg BW per day *(11)*.

Although present concentrations of PCDDs and PCDFs in ambient air do not present a health hazard through direct human exposure, these concentrations will lead to deposition of PCDDs and PCDFs followed by uptake through the food chain.

Health risk evaluation

In 1990, WHO established a tolerable daily intake (TDI) for TCDD of 10 pg/kg BW *(12)*. This was based on TCDD-induced liver cancer in rats *(13)* for which the NOAEL was 1 ng/kg BW. Owing to toxicokinetic differences between humans and rats, this corresponded to a daily intake in humans of 100 pg/kg BW, to which value an uncertainty factor of 10 (to cover inter-individual variation) was applied.

Since then, new data on hormonal, reproductive and developmental effects at low doses in animal studies (rats and monkeys) have been published, and the health risk of dioxins was therefore reassessed in 1998 *(14, 15)*. It was concluded that the human data do not lend themselves to be used as the basis for setting a TDI, but they were nevertheless considered to constitute an important reference for comparison with a health risk assessment based on animal data. Consequently, the TDI was based on animal data. It was further decided that body burdens should be used to scale doses across species. Human daily intakes corresponding to body burdens similar to those associated with LOAELS in rats and monkeys could be estimated to be in the range of 14–37 pg/kg BW per day. By applying an uncertainty factor of 10 to this range of LOAELs, a TDI expressed as a range of 1–4 pg toxic equivalent per kg BW was established for dioxins and dioxin-like compounds.

The TDI represents a tolerable daily intake for lifetime exposure, and occasional short-term excursions above the TDI would have no health consequences provided that the averaged intake over long periods was not exceeded. Although not explicitly stated, the TDI can be looked on as applicable to the total intake of toxic equivalents, via both the oral and inhalation routes, derived from PCDDs and PCDFs and other dioxin-like compounds that act by the same mechanisms and cause similar types of toxicity.

The average daily intake by all routes of exposure to PCDDs and PCDFs, calculated as toxic equivalents, is in the same range as the current TDI. When the contribution from dioxin-like PCBs is taken into account, the intake increases by a factor of 2–3. There are, however, groups with specific dietary habits (such as a high intake of contaminated food) or occupational exposure, that may have exposures in excess of the TDI for PCDDs and PCDFs.

The daily intake of PCDDs and PCDFs in breastfed infants in industrialized countries has been calculated in toxic equivalents to range from about 20 pg/kg BW in less industrialized areas up to about 130 pg/kg BW in more industrialized areas. When the contribution from dioxin-like PCBs is taken into account, the intakes may be up to twice these figures. This indicates intakes being far above the TDI. WHO noted, however, that the TDI should not be applied to breastfed infants because the concept of TDI relates to a dose ingested throughout a lifetime *(14)*. In general, the quantity of PCDDs and PCDFs ingested over a 6-month breastfeeding period would be less than 5% of the quantity ingested over a lifetime.

The contribution from inhalation exposure is on average approximately 1% of the dietary intake, but may in certain extreme situations (areas close to point emission sources or contaminated indoor air) approach the dietary intake.

Guidelines

An air quality guideline for PCDDs and PCDFs is not proposed because direct inhalation exposures constitute only a small proportion of the total exposure, generally less than 5% of the daily intake from food.

Urban ambient toxic equivalent air concentrations of PCDDs and PCDFs are estimated to be about 0.1 pg/m^3. However, large variations have been measured. Although such an air concentration is only a minor contributor to direct human exposure, it is a major contributor to contamination of the food chain. It is difficult, however, to calculate indirect exposure from contamination of food via deposition from ambient air. Mathematical models are being used in the absence of experimental data, but these models require validation. Air concentrations of 0.3 pg/m^3 or higher are indications of local emission sources that need to be identified and controlled.

Although indoor air levels of PCDDs and PCDFs are generally very low, in certain instances, toxic equivalent levels of up to 3 pg/m^3 have been detected. Such levels will constitute an exposure ranging from 25% up to 100% of the current TDI of 1–4 pg toxic equivalent per kg BW (corresponding to 60–240 pg toxic equivalent per day for a 60-kg person).

Owing to the potential importance of the indirect contribution of PCDDs and PCDFs in air to the total human exposure to these compounds through deposition and uptake in the food chain, measures should be undertaken to further reduce emissions to air from known sources. For risk reduction, it is important to control known sources as well as to identify new sources.

References

1. BECK, H. ET AL. PCDD and PCDF body burden from food intake in the Federal Republic of Germany. *Chemosphere,* **18**: 417–424 (1989).
2. MINISTRY OF AGRICULTURE FISHERIES AND FOOD. *Dioxins in food.* London, H. M. Stationery Office, 1992 (Food Surveillance Paper, No. 31).

3. THEELEN, R.M.C. Modeling of human exposure to TCDD and I–TEQ in the Netherlands: background and occupational. *In:* Gallo, M.A. et al., ed. *Biological basis for risk assessment of dioxins and related compounds.* New York, Cold Spring Harbor Laboratory Press, 1991, pp. 277–290 (Banbury Report, No. 35).
4. FÆRDEN, K. *Dioksiner i næringsmidler.* [Dioxins in food]. Oslo, National Food Control Authority, 1991 (SNT-Rapport No. 4).
5. SVENSON, B.G. ET AL. Exposure to dioxins and dibenzofurans through consumption of fish. *New England journal of medicine,* **324**: 8–12 (1991).
6. *Health assessment document for 2,3,7,8–tetrachlorodibenzo*-p-*dioxin (TCDD) and related compounds.* Washington, DC, US Environmental Protection Agency, 1994 (Final report EPA-600/BP-92-001c).
7. AHLBORG, U.G. ET AL. Impact of polychlorinated dibenzo-*p*-dioxins, dibenzofurans and biphenyls on human and environmental health, with special emphasis on application of the toxic equivalency factor concept. *European journal of pharmacology,* **228**: 179–199 (1992).
8. *International toxicity equivalency factors (I–TEF) method of risk assessment for complex mixtures of dioxins and related compounds.* Brussels, North Atlantic Treaty Organization, 1988 (Report No. 176).
9. WEVERS, M. ET AL. Concentrations of PCDDs and PCDFs in ambient air at selected locations in Flanders. *Organohalogen compounds,* **12**: 123–126 (1993).
10. DUARTE-DAVIDSON, R. ET AL. Polychlorinated dibenzo-*p*-dioxins (PCDDs) and furans (PCDFs) in urban air and deposition. *Environmental science and pollution research,* **1**: 262–270 (1994).
11. BALFANZ, E. ET AL. Sampling and analysis – polychlorinated biphenyls (PCB) in indoor air due to permanently elastic sealants. *Chemosphere,* **26**: 871–880 (1993).
12. AHLBORG, U.G. ET AL., ED. Special issue: tolerable daily intake of PCDDs and PCDFs. *Toxic substances journal,* **12**: 101–131 (1992).
13. KOCIBA, R.J. ET AL. Results of a two year chronic toxicity and oncogenicity study of 2,3,7,8-tetrachlordibenzo-*p*-dioxin (TCDD) in rats. *Toxicology and applied pharmacology,* **46**: 279–303 (1978).
14. VAN LEEUWEN, F.X.R. & YOUNES, M.M. Assessment of the health risk of dioxins: re-evaluation of the tolerable daily intake (TDI). *Food additives and contaminants,* **17**: 223–240 (2000).
15. VAN LEEUWEN, F.X.R. ET AL. Dioxins: WHO's Tolerable Daily Intake (TDI) revisited. *Chemosphere,* **40**: 1095–1101 (2000).

5.12 Styrene

Exposure evaluation

Concentrations of styrene in rural ambient air are generally less than 1 μg/m^3, while indoor air in such locations may contain several μg/m^3. Levels in polluted urban areas are generally less than 20 μg/m^3 but can be much higher in newly built houses containing styrene-based materials.

Health risk evaluation

Potentially critical effects for the derivation of a guideline for styrene are considered to be carcinogenicity/genotoxicity and neurological effects, including effects on development.

Styrene in its pure form has an odour detection threshold of 70 μg/m^3. Its pungent odour is recognized at concentrations three to four times greater than this threshold value.

The value of the available evidence for an association between exposure to styrene and small increases in lymphatic and haematopoietic cancers observed in workers in some studies is limited by concurrent exposure to other substances, lack of specificity and absence of dose–response. In limited studies in animals, there is little evidence that styrene is carcinogenic. IARC has classified styrene in group 2B *(1)*.

Styrene was genotoxic *in vivo* and *in vitro* following metabolic activation. In cytogenetic studies on peripheral lymphocytes of workers in the reinforced plastics industry, there were increased rates of chromosomal aberrations at mean levels of styrene of more than 120 mg/m^3 (> 20 ppm). Elevated levels of single-strand breaks and styrene-7,8-oxide adducts in DNA and haemoglobin have also been observed. Although these genotoxic effects have been observed at relatively low concentrations, they were not considered as critical endpoints for development of a guideline, in view of the equivocal evidence of carcinogenicity for styrene.

The available data, although limited, indicate that neurotoxicity in the form of neurological developmental impairments is among the most sensitive of endpoints. In the offspring of rats exposed to styrene at a concentration of 260 mg/m^3 (60 ppm) there were effects on behaviour and biochemical parameters in the brain *(2)*.

Guidelines

Although genotoxic effects in humans have been observed at relatively low concentrations, they were not considered as critical endpoints for development of a guideline, in view of the equivocal evidence for the carcinogenicity of styrene.

In occupationally exposed populations, subtle effects such as reductions in visuomotor accuracy and verbal learning skills *(3–5)* and subclinical effects on colour vision have been observed at concentrations as low as 107–213 mg/m^3 (25–50 ppm) *(6–10)*. Taking the lower number of this range for precautionary reasons, adjusting this to allow for conversion from an occupational to a continuous pattern of exposure (a factor of 4.2), and incorporating a factor of 10 for inter-individual variation and 10 for use of a LOAEL rather than a NOAEL results in a guideline of 0.26 mg/m^3 (weekly average). This value should also be protective for the developmental neurological effects observed in animal species.

The air quality guideline could also be based on the odour threshold. In that case, the peak concentration of styrene in air should be kept below the odour detection threshold level of 70 µg/m^3 as a 30-minute average.

References

1. *Some industrial chemicals*. Lyons, International Agency for Research on Cancer, 1994, pp. 233–320 (IARC Monographs on the Evaluation of Carcinogenic Risks to Humans, Vol. 60).
2. KISHI, R. ET AL. Neurochemical effects in rats following gestational exposure to styrene. *Toxicology letters*, **63**: 141–146 (1992).
3. HÄRKÖNEN, H. Exposure–response relationship between styrene exposure and central nervous function. *Scandinavian journal of work, environment and health*, **4**: 53–59 (1978).
4. LINDSTRÖM, K. ET AL. Relationships between changes in psychological performances and styrene exposure in work with reinforced plastics. *Työ ja ihminen*, **6**: 181–194 (1992) [Swedish].
5. MUTTI, A. ET AL. Exposure–effect and exposure–response relationships between occupational exposure to styrene and neuropsychological functions. *American journal of industrial medicine*, **5**: 275–286 (1984).
6. GOBBA, F. ET AL. Acquired dyschromatopsia among styrene-exposed workers. *Journal of occupational medicine*, **33**: 761–765 (1991).
7. FALLAS, C. ET AL. Subclinical impairment of colour vision among workers exposed to styrene. *British journal of industrial medicine*, **49**: 679–682 (1992).

8. GOBBA, F. & CAVALLERI, A. Kinetics of urinary excretion and effects on colour vision after exposure to styrene. *In:* Sorsa, M. et al., ed. *Butadiene and styrene: assessment of health hazards.* Lyons, International Agency for Research on Cancer, 1993 (IARC Scientific Publication, No. 127), pp. 79–88.
9. CHIA, S.E. ET AL. Impairment of colour vision among workers exposed to low concentrations of styrene. *American journal of industrial medicine,* **26**: 481–488 (1994).
10. EGUCHI, T. ET AL. Impaired colour discrimination among workers exposed to styrene: relevance to a urinary metabolite. *Occupational and environmental medicine,* **52**: 534–538 (1995).

5.13 Tetrachloroethylene

Exposure evaluation

Ambient air concentrations of tetrachloroethylene are generally less than 5 μg/m^3 in urban areas and typically less than 1 μg/m^3 in rural areas. Indoor concentrations are generally less than 5 μg/m^3. Indoor tetrachloroethylene air levels may rise to more than 1 mg/m^3 in close proximity to dry-cleaning operations where tetrachloroethylene is used as a cleaning solvent or in homes where dry-cleaned clothing is often worn. Inhalation of tetrachloroethylene is the major route of exposure in the general population.

Health risk evaluation

The main health effects of concern are cancer and effects on the central nervous system, liver and kidneys. Tetrachloroethylene is classified by IARC as a Group 2A carcinogen (probably carcinogenic to humans) *(1)*.

In carcinogenicity studies, an increased incidence of adenomas and carcinomas was observed in the livers of exposed mice. There is suggestive evidence from mechanistic studies that humans are less sensitive to the development of these tumours following tetrachloroethylene exposure. A low incidence of kidney tumours has been reported among male rats. It can be concluded from this small and statistically non-significant increase, together with the data related to a possible mechanism of induction, that the result in male rats is equivocal evidence only for a risk of renal cancer in humans. The significance for humans of the increased incidences of mononuclear-cell leukaemias, as observed in a study in F344 rats, is unclear owing to the lack of understanding of the mechanism underlying the formation of this cancer type, which has a high background incidence.

Epidemiological studies in humans show positive associations between exposure to tetrachloroethylene and risks for oesophageal and cervical cancer and non-Hodgkin lymphoma. Confounding factors cannot be ruled out and the statistical power of the studies is limited. These studies therefore provide only limited evidence for the carcinogenicity of tetrachloroethylene in humans *(1)*.

From the weight of the evidence from mutagenicity studies, it can be concluded that tetrachloroethylene is not genotoxic. Several *in vitro* studies indicate that conjugation of tetrachloroethylene with reduced glutathione, a minor biotransformation route demonstrated to occur in rodents, produces renal metabolites that are mutagenic in *Salmonella typhimurium* TA 100 *(1)*. In the absence of further data on this point, the significance of the latter results for humans is uncertain.

Short-term exposure studies in volunteers (duration 1 or 5 days) have shown effects on the central nervous system at a concentration of > 678 mg/m^3 *(2–5)*. A recent study of dry-cleaning workers with long-term exposure showed that renal effects may develop at lower exposure concentrations, with the reported onset of renal damage occurring following exposure to a median concentration of 102 mg/m^3 (range, trace to 576 mg/m^3) *(6)*.

Although the results of carcinogenicity studies in experimental animals are available, those of adequate long-term toxicity studies are not. A chronic LOAEL of 678 mg/m^3 (100 ppm) for the systemic toxicity (in kidney and liver) of tetrachloroethylene in mice can be derived from the National Toxicology Program carcinogenicity study in this species *(7)*.

Use of existing physiologically based pharmacokinetic models for derivation of a guideline value based on kidney effects is not considered feasible because these models do not contain the kidney or kidney-specific metabolism as a component. As yet it is therefore unknown what an appropriate internal dose measure would be.

Guidelines

Given the limitations of the weight of the epidemiological evidence, and the uncertainty of the relevance to humans of the induction of tumours in animals exposed to tetrachloroethylene, the derivation of a guideline value is at present based on non-neoplastic effects rather than on carcinogenicity as the critical endpoint.

On the basis of a long-term LOAEL for kidney effects of 102 mg/m^3 in dry-cleaning workers, a guideline value of 0.25 mg/m^3 is calculated. In deriving this guideline value, the LOAEL is converted to continuous exposure (dividing by a factor of 4.2, 168/40) and divided by an uncertainty factor of 100 (10 for use of an LOAEL and 10 for intraspecies variation). Recognizing that some uncertainty in the LOAEL exists because the effects observed at this level are not clear-cut, and because of fluctuations in exposure levels, an alternative calculation was made based on the LOAEL in mice of 680 mg/m^3, and using an appropriate uncertainty factor of 1000. This calculation yields a guideline value of 0.68 mg/m^3.

On the basis of the overall health risk evaluation, a guideline of 0.25 mg/m^3 is currently established. However, the concern about a possible carcinogenic effect of tetrachloroethylene exposure in humans should be addressed through in-depth risk evaluation in the near future.

References

1. *Dry cleaning, some chlorinated solvents and other industrial chemicals.* Lyons, International Agency for Research on Cancer, 1995 (IARC Monographs on the Evaluation of Carcinogenic Risks to Humans, Vol. 63).
2. *Toxicological profile for tetrachloroethylene.* Atlanta, GA, Agency for Toxic Substances and Disease Registry, 1993 (Public Health Service Report, No. TP-92/18).
3. *Tetrachloroethylene.* Geneva, World Health Organization, 1984 (Environmental Health Criteria, No. 31).
4. HAKE, C.L. & STEWART, R.D. Human exposure to tetrachloroethylene: inhalation and skin contact. *Environmental health perspectives,* **21**: 231–238 (1977).
5. ALTMANN, L. ET AL. Neurophysiological and psychological measurements reveal effects of acute low-level organic solvent exposure in humans. *International archives of occupational and environmental health,* **62**: 493–499 (1990).
6. MUTTI, A. ET AL. Nephropathies and exposure to perchloroethylene in dry-cleaners. *Lancet,* **340**: 189–193 (1992).
7. *Technical report on the toxicology carcinogenesis studies of tetrachloroethylene (perchloroethylene) in F344/N rats and B6C3F1 mice (inhalation studies).* Research Triangle Park, NC, National Toxicology Program, 1986 (NTP Technical Report, No. 311).

5.14 Toluene

Exposure evaluation

Mean ambient air concentrations of toluene in rural areas are generally less than 5 μg/m^3, while urban air concentrations are in the range 5–150 μg/m^3. Concentrations may be higher close to industrial emission sources.

Health risk evaluation

Toluene in its pure form has an odour detection threshold of 1 mg/m^3 *(1, 2)*. Its odour is recognized at concentrations about ten times greater than this threshold value *(1–3)*.

The acute and chronic effects of toluene on the central nervous system are the effects of most concern. Toluene may also cause developmental decrements and congenital anomalies in humans, and these effects are supported by findings of studies on animals, for example fetal development retardation, skeletal anomalies, low birth weight and developmental neurotoxicity. The potential effects of toluene on reproduction and hormone balance in women, coupled with findings of hormone imbalances in exposed males, are also of concern. Limited information suggests an association between occupational toluene exposure and spontaneous abortions. Both the human and animal data indicate that toluene is ototoxic at elevated exposures. Sensory effects have also been found. Toluene has minimal effects on the liver and kidney, except in cases of toluene abuse. There has been no indication that toluene is carcinogenic in bioassays conducted to date, and the weight of available evidence indicates that it is not genotoxic.

The lowest level of chronic occupational toluene exposure unequivocally associated with neurobehavioural functional decrements is 332 mg/m^3 (88 ppm) *(4, 5)*. Effects on the central nervous system in humans are supported by findings in exposed animals. For example, rat pups exposed to either 100 or 500 ppm 1–28 days after birth demonstrated histo-pathological changes in the hippocampus *(6)*. Women occupationally exposed to toluene at an average concentration of 332 mg/m^3 (88 ppm) incurred higher spontaneous abortion rates and menstrual function disturbances *(7–9)*. The interpretation of these observations was hampered, however, by confounding factors *(10)*. Men occupationally exposed to toluene at 5–25 ppm have also been shown to exhibit hormonal changes.

With regard to short-term exposure, subjective effects have been reported at 100 ppm (6-hour exposure) while symptoms at lower levels cannot be ruled out. Numerous confounding factors, however, need to be considered.

Exposure data related to central nervous system endpoints were best characterized in certain occupational studies and these data have been employed in the derivation of the guideline. A NOAEL for chronic effects of toluene has not been identified.

Guidelines

The LOAEL for effects on the central nervous system from occupational studies is approximately 332 mg/m^3 (88 ppm). A guideline value of 0.26 mg/m^3 is established from these data, adjusting for continuous exposure (dividing by a factor of 4.2) and dividing by an uncertainty factor of 300 (10 for inter-individual variation, 10 for use of a LOAEL rather than a NOAEL, and an additional factor of 3 given the potential effects on the developing central nervous system). This guideline value should be applied as a weekly average. This guideline value should also be protective for reproductive effects (spontaneous abortions).

The air quality guideline could also be based on the odour threshold. In this case, the peak concentrations of toluene in air should be kept below the odour detection threshold level of 1 mg/m^3 as a 30-minute average.

References

1. Hellman, T.M. & Small, F.H. Characterization of petrochemical odors. *Chemical engineering progress*, **69**: 75–77 (1973).
2. Nauš, A. Cichové prahy nekterých prumyslových látek [Olfactory thresholds of industrial substances]. *Pracovni lekarstvi*, **34**: 217–218 (1982).
3. Hellman, T.M. & Small, F.H. Characterization of the odor properties of 101 petrochemicals using sensory methods. *Journal of the Air Pollution Control Association*, **24**: 979–982 (1974).
4. Foo, S.C. et al. Chronic neurobehavioural effects of toluene. *British journal of industrial medicine*, **47**: 480–484 (1990).
5. Foo, S.C. et al. Neurobehavioural effects in occupational chemical exposure. *Environmental research*, **60**: 267–273 (1993).
6. Slomianka, L. et al. The effect of low-level toluene exposure on the developing hippocampal region of the rat: histological evidence and volumetric findings. *Toxicology*, **62**: 189–202 (1990).
7. Ng, T.P. et al. Risk of spontaneous abortion in workers exposed to toluene. *British journal of industrial medicine*, **49**: 804–808 (1992).

8. LINDBOHM, M.-L. ET AL. Spontaneous abortions among women exposed to organic solvents. *American journal of industrial medicine,* **17**: 449–463 (1990).
9. NG, T.P. ET AL. Menstrual function in workers exposed to toluene. *British journal of industrial medicine,* **49**: 799–803 (1992).
10. *Air quality guidelines for Europe.* Copenhagen, WHO Regional Office for Europe, 1987 (WHO Regional Publications, European Series, No. 23).

5.15 Trichloroethylene

Exposure evaluation

The average ambient air concentrations of trichloroethylene are less than 1 μg/m^3 in rural areas and up to 10 μg/m^3 in urban areas. Concentrations in indoor air are typically similar, although higher concentrations can be expected in certain areas, such as in proximity to industrial operations. Inhalation of airborne trichloroethylene is the major route of exposure for the general population.

Health risk evaluation

The main health effects of concern with trichloroethylene are cancer, and effects on the liver and the central nervous system.

Studies in animals and humans show that the critical organs or systems for noncarcinogenic effects are the liver and the central nervous system. The dose–response relationship for these effects is insufficiently known, making it difficult to assess the health risk for the occurrence of these effects in case of long-term exposure to low levels of trichloroethylene.

IARC has classified trichloroethylene as a Group 2A carcinogen (probably carcinogenic to humans). This classification was based on sufficient evidence in animals and limited evidence in humans *(1)*.

The available data suggest that trichloroethylene may have a weak genotoxic action *in vivo*. Several of the animal carcinogenicity studies show limitations in design. In mice, increased incidences of adenomas and carcinomas in lungs and liver were observed *(2–5)*. In two rat studies, the incidence of testicular tumours was increased *(6, 7)*. Evidence from mechanistic studies suggests that humans are likely to be less susceptible to developing tumours as a result of exposure to trichloroethylene. Nevertheless, the relevance of the observed increase in lung tumours in mice and testicular tumours in rats for human cancer risks cannot be excluded. The results of the mechanistic studies do not provide full elucidation or guidance on this point.

Positive associations between exposure to trichloroethylene and risks for cancer of the liver and biliary tract and non-Hodgkin lymphomas were observed in epidemiological studies on cancer in humans. Confounding cannot be ruled out. A quantitative risk estimate cannot be made from these human data. The increased tumours in lungs and testes observed in animal bioassays are considered to be the best available basis for the risk evaluation. However, it cannot be conclusively established whether a threshold with regard to carcinogenicity in the action of trichloroethylene may be assumed. Therefore, linear extrapolation from the animal tumour data is used, providing a conservative approach to the estimation of human cancer risk.

Using the data on the incidence of pulmonary adenomas in B3C6F1 mice and on pulmonary adenomas/carcinomas in Swiss mice *(2)*, unit risks of 9.3×10^{-8} and 1.6×10^{-7}, respectively, can be calculated by applying the linearized multistage model. Applying the same model on the incidence of Leydig cell tumours in the testes of rats, a unit risk of 4.3×10^{-7} can be derived *(6)*.

Physiologically based pharmacokinetic models have been developed for trichloroethylene. Use of these models for cancer risk estimates is not considered feasible because it is not known what an appropriate internal dose measure would be.

Guidelines

Because the available evidence indicates that trichloroethylene is genotoxic and carcinogenic, no safe level can be recommended. On the basis of the most sensitive endpoint, Leydig cell tumours in rats, a unit risk estimate of 4.3×10^{-7} per μg/m³ can be derived. The ranges of ambient air concentrations of trichloroethylene corresponding to an excess lifetime risk of 1:10 000, 1:100 000 and 1:1 000 000 are 230, 23 and 2.3 μg/m³, respectively.

References

1. *Dry cleaning, some chlorinated solvents and other industrial chemicals*. Lyons, International Agency for Research on Cancer, 1995 (IARC Monographs on the Evaluation of Carcinogenic Risks to Humans, Vol. 63).
2. Maltoni, C. et al. Long-term carcinogenicity bioassays on trichloroethylene administered by inhalation to Sprague-Dawley rats and Swiss mice and $B3C6F_1$ mice. *Annals of the New York Academy of Science*, **534**: 316–342 (1988).
3. Fukuda, K. et al. Inhalation carcinogenicity of trichloroethylene in mice and rats. *Industrial health*, **21**: 243–254 (1983).
4. *Technical report on the carcinogenesis bioassay of trichloroethylene, CAS No. 79–01–6*. Bethesda, MD, US Department of Health, Education and Welfare, 1976 (DHEW Publication, No. 76–802; National Cancer Institute Technical Report, No. 2).

5. *Technical report on the carcinogenesis studies of trichloroethylene (without epichlorohydrin) in F344/N rats and B6C3F1 mice (gavage studies)*. Research Triangle Park, NC, National Toxicology Program, 1990 (NTP Technical Report, No. 243).
6. MALTONI, C. ET AL. Experimental research on trichloroethylene carcinogenesis. *In:* Maltoni, C. & Mehlman, M.A., ed. *Archives of research on industrial carcinogenesis 5*. Princeton, NJ, Princeton Science Publishers, 1986.
7. *Technical report on the carcinogenesis studies of trichloroethylene in four strains of rats (ACI, August, Marshall, Osborne-Mendel) (gavage studies)*. Research Triangle Park, NC, National Toxicology Program, 1988 (NTP Technical Report, No. 273).

5.16 Vinyl chloride

Exposure evaluation

Calculations based on dispersion models indicate that 24-hour average concentrations of 0.1–0.5 μg/m^3 exist as background levels in much of western Europe, but such concentrations are below the current detection limit (approximately 1.0 μg/m^3). In the vicinity of vinyl chloride (VC) and polyvinyl chloride (PVC) production facilities 24-hour concentrations can exceed 100 μg/m^3, but are generally less than 10 μg/m^3 at distances greater than 1 km from plants. The half-time of VC in the air is calculated to be 20 hours; this figure is based on measured rates of reaction with hydroxyl radicals and their concentrations in the air *(1)*.

Health risk evaluation

There is sufficient evidence of carcinogenicity of VC in humans and experimental animals *(2)*. Extrapolation (or rather interpolation) to lower exposure levels can be made, based on knowledge or assumptions about the dose and time-dependence of risk. As seen in the low exposure data of Maltoni et al. *(3)*, a linear dose–response relationship accords well with the animal data for haemangiosarcoma. The finding of at least three cases of haemangiosarcoma in PVC processors as compared with about 100 in VC or PVC production workers is compatible with a linear relationship. The average exposures in the production industry were about 100 times lower than those in the polymerization industry, but the workforce was 10 times larger.

Data from a cohort study *(4)* and an analysis of the incidence of haemangiosarcoma in the United States and western Europe *(5)* suggest that the risk of haemangiosarcoma increases as the second or third power of time from onset of exposure. Using a model in which the risk increases as t^3 during exposure and as t^2 subsequently, estimates of the relative risk in various exposure circumstances can be calculated and used to convert limited-duration exposure risks into lifetime exposure risks.

Estimates of cancer risk can be made from the data relating to the cohort studied by Nicholson et al. *(4)*. A group of 491 workers at two long-established PVC production plants was studied. One plant began operations in 1936 and the other in 1946. Each cohort member had a minimum of 5 years' employment; the average work duration was 18 years. It is estimated that the average VC exposure was 2050 mg/m^3. The overall standardized mortality rate (SMR) for cancer was 142 (28 observed; 19.7 expected); that for liver and biliary cancer was 2380 (10 observed; 0.42 expected). Using the liver cancer data, the estimated lifetime risk of death from VC exposure is 3.6×10^{-4} per mg/m^3, or [(23.8–1) × 0.003/ (2050 mg/m^3) × 2.8 × 70/18], where 0.003 is the lifetime risk of death from liver biliary cancer in white American males, 2.8 is the working week–total week conversion and 70/18 the work period–lifetime conversion. Since there are an equal number of cancers at other sites (averaging over 12 cohorts), the excess cancer risk is 7.2×10^{-4} per mg/m^3. If the total cancer SMR is used directly, the risk is 4.5×10^{-4} per mg/m^3, or [(1.42–1) × 0.2/ (2050 mg/m^3) × 2.8 × 70/18], which is in good agreement with the above. The average of the two estimates indicates that a 10^{-6} cancer risk occurs at exposures of 1.7 µg/m^3.

The risk of cancer from VC can be calculated from data on the United States population exposed in the Equitable Environmental Health study *(6)*. This study identified 10 173 workers who were employed for one or more years in 37 (of 43) VC and PVC production plants. The average duration of employment before 1973 was 8.7 years. Using the data of Barnes *(7)*, a weighted exposure of 650 ppm (1665 mg/m^3) was estimated. Considering the total population at risk to be 12 000, the unit exposure lifetime risk from an average exposure of 9 years is 0.75×10^{-5} per mg/m^3, or [(150/12 000) × (1/1665)].

Using a linear dose–response relationship converting to a lifetime exposure (assuming that one half of the workers began exposure at the age of 20 and one half at the age of 30), the continuous lifetime haemangiosarcoma risk is 4.7×10^{-4} per mg/m^3, or [$0.75 \times 10^{-5} \times 2.8 \times 22.4$], where 2.8 is the ratio of the air volume inhaled in a full week (20 m^3 × 7) to that in a working week (10 m^3 × 5) and 22.4 is the average conversion to a lifetime for a ten-year exposure beginning at an average age of 25 years, taking into account the time course of haemangiosarcoma. (Without explicit consideration of the time course, the multiplier would be 70/9 = 7.8.) A 10^{-6} risk occurs at a concentration of 2.1 µg/m^3.

Assuming that the number of cancers in other sites may equal that of haemangiosarcomas, the best estimate for excess cancer risk is that a 10^{-6} risk occurs as a result of continuous lifetime exposure to 1.0 µg/m^3.

The risks estimated from epidemiological studies are the most relevant for human exposures. The above estimate from human angiosarcoma incidences is a conservative one from the point of view of health, because of the use of a model that assumes that the haemangiosarcoma risk continues to increase throughout the lifetime of an exposed individual.

These risk estimates are in agreement with those made by others. The US Environmental Protection Agency has estimated that 11 cancer deaths per year would result from 4.6×10^{-6} people being exposed to 0.017 ppm (43 $\mu g/m^3$) *(8)*: this translates to a 10^{-6} lifetime risk at 0.25 $\mu g/m^3$. A Dutch criteria document, on the basis of animal data, estimates that a 10^{-6} risk occurs at 0.035 $\mu g/m^3$ *(1)*.

One cautionary note should be sounded: the particular sensitivity of newborn rats to VC, referred to above, suggests that risks may be much greater in childhood than those estimated from adult exposures. By the age of 10 years, however, the latter risks should prevail.

Guidelines

Vinyl chloride is a human carcinogen and the critical concern with regard to environmental exposures to VC is the risk of malignancy. No safe level can be indicated. Estimates based on human studies indicate a lifetime risk from exposure to 1 $\mu g/m^3$ to be 1×10^{-6}.

References

1. *Criteriadocument over vinylchloride* [Vinyl chloride criteria document]. The Hague, Ministry of Housing, Spatial Planning and Environment, 1984 (Publikatiereeks Lucht, No. 34).
2. *Some monomers, plastics and synthetic elastomers, and acrolein.* Lyons, International Agency for Research on Cancer, 1979 (IARC Monographs on the Evaluation of the Carcinogenic Risk of Chemicals to Humans, Vol. 19).
3. MALTONI, C. ET AL. Carcinogenicity bioassays of vinyl chloride monomer: a model of risk assessment on an experimental basis. *Environmental health perspectives,* **41**: 3–29 (1981).
4. NICHOLSON, W.J. ET AL. Occupational hazards in the VC-PVC industry. *In:* Jarvisalo, P. et al., ed. *Industrial hazards of plastics and synthetic elastomers.* New York, Alan R. Liss, 1984 (Progress in Clinical and Biological Research, Vol. 141), pp. 155–176.
5. NICHOLSON, W.J. ET AL. Trends in cancer mortality among workers in the synthetic polymers industry. *In:* Jarvisalo, P. et al., ed. *Industrial hazards of plastics and synthetic elastomers.* New York, Alan R. Liss, 1984, pp. 65–78 (Progress in Clinical and Biological Research, Vol. 141).
6. *Epidemiological study of vinyl chloride workers.* Rockville, MD, Equitable Environmental Health, Inc., 1978.

7. BARNES, A.W. Vinyl chloride and the production of PVC. *Proceedings of the Royal Society of Medicine,* **69**: 277–281 (1976).
8. KUZMACH, A.M. & MCGAUGHY, R.E. *Quantitative risk assessment for community exposure to vinyl chloride.* Washington, DC, US Environmental Protection Agency, 1975.

CHAPTER 6

Inorganic pollutants

6.1 Arsenic

Exposure evaluation

There are many arsenic compounds, both organic and inorganic, in the environment. Airborne concentrations of arsenic range from 1 ng/m^3 to 10 ng/m^3 in rural areas and from a few nanograms per cubic metre to about 30 ng/m^3 in noncontaminated urban areas. Near emission sources, such as nonferrous metal smelters and power plants burning arsenic-rich coal, concentrations of airborne arsenic can exceed 1 μg/m^3.

Health risk evaluation

Inorganic arsenic can have acute, subacute and chronic effects, which may be either local or systemic. Lung cancer is considered to be the critical effect following inhalation. An increased incidence of lung cancer has been seen in several occupational groups exposed to inorganic arsenic compounds. Some studies also show that populations near emission sources of inorganic arsenic, such as smelters, have a moderately elevated risk of lung cancer. Information on the carcinogenicity of arsenic compounds in experimental animals was considered inadequate to make an evaluation *(1, 2)*.

A significant number of studies concerning occupational exposure to arsenic and the occurrence of cancer have been described. Unit risks derived by the US Environmental Protection Agency (EPA) Carcinogen Assessment Group in 1984 *(3)* were not changed until 1994 *(4)*. They form five sets of data involving two independently exposed populations of workers in Montana and Tacoma smelters in the United States, ranging from 1.25×10^{-3} to 7.6×10^{-3}, a weighted average of these five estimates giving a composite estimate of 4.29×10^{-3}.

A WHO Working Group on Arsenic *(5)* conducted a quantitative risk assessment, assuming a linear relationship between the cumulative arsenic dose and the relative risk of developing lung cancer. Risk estimates for lung cancer from inorganic arsenic exposure were based on the study by Pinto et al. *(6)* of workers at the Tacoma smelter. The lifetime risk of lung cancer was calculated to be 7.5×10^{-3} per microgram of airborne arsenic per cubic metre.

The second study relating to the quantitative risk assessment included a large number of the 8047 males employed as smelting workers at the Montana copper smelter *(7)*. Exposures to airborne arsenic levels were estimated to average 11.17, 0.58 and 0.27 mg/m^3 in the high-, medium- and low-exposure areas. Unit risks for these three groups were calculated to be 3.9×10^{-3}, 5.1×10^{-3} and 3.1×10^{-3}, respectively.

Assuming that the risk estimation based on the Tacoma study was higher because of the urine measurements made, it may have underestimated the actual inhalation exposure; the unit risk was considered to be 4×10^{-3}.

In 1994, Viren & Silvers *(8)*, using updated results from the cohort mortality study in the Tacoma smelter workers together with findings from a cohort study of 3619 Swedish smelter workers, developed other unit risk estimates. A unit risk of 1.28×10^{-3} was estimated for the Tacoma smelter cohort and 0.89×10^{-3} for the Swedish cohort. Pooling these new estimates with the EPA's earlier estimates from the Montana smelter yielded a composite unit risk of 1.43×10^{-3} (Table 13). This value is three times lower than the EPA estimate *(4)* and two times lower than the value assumed in the first edition of *Air quality guidelines for Europe (9)*.

Table 13. Updated unit risk estimates

Risk update	Smelter population	Estimated unit risk: Study	Cohort	Pooled
Pooled estimate using updated Swedish and Tacoma cohorts	Tacoma, 1987	1.28×10^{-3}	1.28×10^{-3}	1.07×10^{-3}
	Ronnskar, 1989:			
	– workers hired pre-1940	0.46×10^{-3}	0.89×10^{-3}	
	– workers hired 1940 and later	1.71×10^{-3}	–	
Updated Tacoma cohort with original EPA estimates for Montana cohort	Tacoma, 1987 (updated results supersede earlier estimates)		1.28×10^{-3}	1.81×10^{-3}
	Montana, 1984 (new estimates not available, 1984 EPA estimates apply)		2.56×10^{-3}	
Pooled across all smelter cohorts	Ronnskar, 1989		0.89×10^{-3}	1.43×10^{-3}
	Tacoma, 1987		1.28×10^{-3}	
	Montana, 1984		2.56×10^{-3}	

Source: Viren & Silvers *(8)*.

Guidelines

Arsenic is a human carcinogen. Present risk estimates have been derived from studies in exposed human populations in Sweden and the United States. When assuming a linear dose–response relationship, a safe level for inhalation exposure cannot be recommended. At an air concentration of 1 $\mu g/m^3$, an estimate of lifetime risk is 1.5×10^{-3}. This means that the excess lifetime risk level is 1:10 000, 1:100 000 or 1:1 000 000 at an air concentration of about 66 ng/m^3, 6.6 ng/m^3 or 0.66 ng/m^3, respectively.

References

1. WOOLSON, E.A. Man's perturbation of the arsenic cycle. *In:* Lederer, W.H. & Fensterheim, R.J., ed. *Arsenic: industrial, biomedical and environmental perspectives. Proceedings of the Arsenic Symposium, Gaithersburg, MD.* New York, Van Nostrand Reinhold, 1983.
2. *Overall evaluations of carcinogenicity: an updating of IARC monographs volumes 1 to 42.* Lyons, International Agency for Research on Cancer, 1987 (IARC Monographs on the Evaluation of the Carcinogenic Risk of Chemicals to Humans, Supplement 7).
3. *Health assessment document for inorganic arsenic.* Washington, DC, US Environmental Protection Agency, 1984 (Final report EPA-600-8-83-021F).
4. INTEGRATED RISK INFORMATION SYSTEM (IRIS). *Carcinogenicity assessment for lifetime exposure to arsenic* (http://www.epa.gov/ngispgm3/iris/subst/0278.htm#I.B). Cincinnati, OH, US Environmental Protection Agency (accessed January 1994).
5. *Arsenic.* Geneva, World Health Organization, 1981 (Environmental Health Criteria, No. 18).
6. PINTO, S.S. ET AL. Mortality experience in relation to a measured arsenic trioxide exposure. *Environmental health perspectives*, **19**: 127–130 (1977).
7. LEE-FELDSTEIN, A. Arsenic and respiratory cancer in man: follow-up of an occupational study. *In:* Lederer, W.H. & Fensterheim, R.J., ed. *Arsenic: industrial, biomedical and environmental perspectives. Proceedings of the Arsenic Symposium, Gaithersburg, MD.* New York, Van Nostrand Reinhold, 1983, pp. 245–254.
8. VIREN, J.R. & SILVERS, A. Unit risk estimates for airborne arsenic exposure: an updated view based on recent data from two copper smelter cohorts. *Regulatory toxicology and pharmacology*, **20**: 125–138 (1994).
9. *Air quality guidelines for Europe.* Copenhagen, WHO Regional Office for Europe, 1987 (WHO Regional Publications, European Series, No. 23).

6.2 Asbestos

Exposure evaluation

Actual indoor and outdoor concentrations in air range from below one hundred to several thousand fibres per m^3.

Health risk evaluation

On the basis of the evidence from both experimental and epidemiological studies, it is clear that asbestos inhalation can cause asbestosis, lung cancer and mesothelioma. The evidence that ingested asbestos causes gastrointestinal or other cancers is insufficient. Furthermore, the carcinogenic properties of asbestos are most probably due to its fibre geometry and remarkable integrity; other fibres with the same characteristics may also be carcinogenic.

Current environmental concentrations of asbestos are not considered a hazard with respect to asbestosis. However, a risk of mesothelioma and lung cancer from the current concentrations cannot be excluded.

In 1986 a WHO Task Group expressed reservations about the reliability of risk assessment models applied to asbestos risk. Its members suggested that such models can only be used to obtain a broad approximation of the lung cancer risk of environmental exposures to asbestos and "that any number generated will carry a variation over many orders of magnitude". The same was found to be true for estimates of the risk of mesothelioma. The same document stated: "In the general population the risks of mesothelioma and lung cancer attributable to asbestos cannot be quantified reliably and probably are undetectably low." *(1)*.

The following estimates of risk are based on the relatively large amount of evidence from epidemiological studies concerning occupational exposure. Data from these studies have been conservatively extrapolated to the much lower concentrations found in the general environment. Although there is evidence that chrysotile is less potent than amphiboles, as a precaution chrysotile has been attributed the same risk in these estimates.

Mesothelioma

A formula by which the excess incidence of mesothelioma can be approximated has been derived by Peto *(2)*. Fibre concentration, duration of exposure and time since first exposure are parameters incorporated in this model, which assumes a linear dose–response relationship. Peto verified this model from data on an urban population exposed for its whole life and on workers exposed for many decades. In both cases, duration of exposure is assumed to be equal or close to time since first exposure. The data show that the incidence of mesothelioma is proportional to the fibre concentration to which the workers were exposed and to time since first exposure for both workers and the general population. Starting from this relationship, one may calculate the risk of lifetime exposure to environmental concentrations from the incidence of mesothelioma in occupational populations exposed to much higher concentrations, but for a shorter time.

Apart from incomplete knowledge about the true workplace exposure, a further complication arises from the fact that workplace concentrations were measured by means of an optical microscope, counting only fibres longer than 5 μm and thicker than, say, 0.5 μm. In this chapter all fibre concentrations based on optical microscopy are marked F^*/m^3 and risk estimates will be based on F^*/m^3. If concentrations measured by optical microscopy are to be compared with environmental fibre concentrations measured by scanning electron microscopy, a conversion factor has to be used: $2\ F/m^3 = 1\ F^*/m^3$.

Several studies have been performed to calculate the risk of mesothelioma resulting from nonoccupational exposure to asbestos. Lifetime exposure to 100 F^*/m^3 has been estimated by various authors to carry differing degrees of mesothelioma risk (see Table 14). The risk estimates in Table 14 differ by a factor of 4. A "best" estimate may be 2×10^{-5} for 100 F^*/m^3.

An independent check of this risk estimate can be made by calculating the incidence of mesothelioma in the general population, based on a hypothetical

Table 14. Estimates of mesothelioma risk resulting from lifetime exposure to asbestos

Risk of mesothelioma from 100 F^*/m^3	Values in original publication (risk for fibre concentration indicated)	Reference
1.0×10^{-5}	1.0×10^{-4} for 1000 F^*/m^3	*(3)*
~2.0×10^{-5}	1.0×10^{-4} for (130–800) F^*/m^3	*(4)*
~3.9×10^{-5}	1.56×10^{-4} for 400 F^*/m^3	*(5, 6)*
~2.4×10^{-5}	2.75×10^{-3} (females), 1.92×10^{-3} (males) for 0.01 F/ml	*(7)*

average asbestos exposure 30–40 years ago *(8)*. If the latter had been 200–500 F*/m^3 (corresponding to about 400–1000 F/m^3 as measured today), the resulting lifetime risk of mesothelioma would be (4–10) × 10^{-5}. With the average United States death rate of 9000 × 10^{-6} per year, this would give 0.4–0.9 mesothelioma cases each year per million persons from past environmental asbestos exposure. The reported mesothelioma incidence in the United States ranges from 1.4 × 10^{-6} per year to 2.5 × 10^{-6} per year according to various authors *(5, 8)*. Thus, the calculated risk figures would account for only part of the observed incidence. Nevertheless, other factors that may account for this discrepancy must be considered.

- Uncertainties in the risk extrapolations result from the lack of reliable exposure data in the cohort studies, errors in the medical reports, and necessary simplifications in the extrapolation model itself *(7)*. Furthermore, the amount of past ambient exposure can only be an educated guess.

- The incidence of nonoccupational mesotheliomas is calculated from the difference between the total of observed cases and the number of those probably related to occupational exposure. Neither of these two figures is exactly known. Moreover, the influence of other environmental factors in the generation of mesothelioma is unknown.

In the light of these uncertainties, the result obtained by using the risk estimate can be considered to be in relatively good agreement with the annual mesothelioma death rate based on national statistical data.

Lung cancer

Unlike mesothelioma, lung cancer is one of the most common forms of cancer. As several exogenous noxious agents can be etiologically responsible for bronchial carcinoma, the extrapolation of risk and comparison between different studies is considerably complicated. In many epidemiological studies, the crucial effect of smoking has not been properly taken into account.

Differentiation of the observed risks according to smoking habits has been carried out, however, in the cohort of North American insulation workers studied by Hammond et al. *(9)*. This study suggests that the relative risk at a given time is approximately proportional to the cumulative amount of fine asbestos dust received up to this point, for both smokers and nonsmokers. The risks for non-asbestos-exposed nonsmokers and smokers must therefore be multiplied by a factor that increases in proportion to the cumulative exposure.

The dose–response relationship in the case of asbestos-induced lung cancer can be described by the following equation *(7)*.

$$I_L \text{ (age, smoking, fibre dose)} = I_L^o \text{ (age, smoking)} [1 + K_L \times C_f \times d]$$

This equation could also be written as:

$$K_L = [(I_L/I_L^o) - 1]/C_f \times d = \text{(relative risk} - 1)/\text{(cumulative exposure)}$$

where:

K_L = a proportionality constant, which is a measure of the carcinogenic potency of asbestos

C_f = fibre concentration

d = duration of exposure in years

I_L = lung cancer incidence, observed or projected, in a population exposed to asbestos concentration C_f during time d

I_L^o = lung cancer incidence expected in a group without asbestos exposure but with the same age and smoking habits (this factor includes age dependence).

There are several studies that allow the calculation of K_L. Liddell *(10, 11)* has done this in an interesting and consistent manner. The results are given in Table 15.

Taking the data in Table 15 as a basis, a reasonable estimate for K_L is 1.0 per 100 F*years/ml. For a given asbestos exposure, the risk for smokers is about 10 times that for nonsmokers *(9)*. In extrapolating from workers to the general public, a factor of 4 for correction of exposure time has to be applied to K_L.

The incidence of lung cancer in the general population exposed to 100 F*/m^3 is calculated as follows:

$$I_L = I_L^o(1 + 4 \times 0.01 \times 10^{-4} \text{ F*/ml} \times 50 \text{ years})$$

or

$$I_L = I_L^o(1 + 2 \times 10^{-4} \text{ F*/ml})$$

Table 15. Increase in the relative risk of lung cancer, as shown by different studies

K_L per 100 F*year/ml	Type of activity	Reference
0.04	mining and milling	*(12)*
0.045	mining and milling	*(13)*
0.06	friction material	*(14)*
0.1	factory processes	*(15, 16)*
(M) 0.4–1.1	factory processes	
(F) 2.7 [a]	factory processes	*(17)* [b]
0.2	asbestos-cement	*(18)*
0.07	textiles (before 1951)	*(19)*
0.8 [a]	textiles (after 1950)	
6(M) 1.6 [a]	textiles	*(20)*
1.6	textiles	*(21)* [c]
1.1	insulation products	*(22)* [b]
1.5	insulation	*(23)* [b]

[a] Fewer than 10 cases of lung cancer expected (i.e. small cohort).

[b] Inadequate knowledge of actual fibre concentrations.

[c] Same factory as in *(20)*, but larger cohort.

Source: Liddell *(10)*.

The extra risk is $I_L - I_L^o$. Values for I_L^o are about 0.1 for male workers and 0.01 for male nonsmokers *(5)*.

Lifetime exposure to 100 F*/m³ (lifetime assumed to be 50 years since, in a lifetime of 70 years, the first 20 years without smoking probably do not make a large contribution) is therefore estimated as follows.

Status	Risk of lung cancer per 100 000	Range (using the highest and lowest values of K_L from Table 15)
Smokers	2.0	0.08–3.2
Nonsmokers	0.2	0.008–0.32

This risk estimate can be compared, when adjusted to 100 F^*/m^3, with estimates for male smokers made by other authors or groups:

Breslow (National Research Council) *(6)*: 7.3×10^{-5}

Schneiderman et al. *(4)*: $(14–1.4) \times 10^{-5}$

US Environmental Protection Agency *(7)*: 2.3×10^{-5}.

A fibre concentration of 100 F^*/m^3 (about 200 F/m^3 as seen by scanning electron microscope) thus gives a total risk of $(2 + 2) \times 10^{-5}$ for smokers or 2.2×10^{-5} for nonsmokers.

Guidelines

Asbestos is a proven human carcinogen (IARC Group 1). No safe level can be proposed for asbestos because a threshold is not known to exist. Exposure should therefore be kept as low as possible.

Several authors and working groups have produced estimates indicating that, with a lifetime exposure to 1000 F/m^3 (0.0005 F^*/ml or 500 F^*/m^3, optically measured) in a population of whom 30% are smokers, the excess risk due to lung cancer would be in the order of $10^{-6}–10^{-5}$. For the same lifetime exposure, the mesothelioma risk for the general population would be in the range $10^{-5}–10^{-4}$. These ranges are proposed with a view to providing adequate health protection, but their validity is difficult to judge. An attempt to calculate a "best" estimate for the lung cancer and mesothelioma risk is described above.

References

1. *Asbestos and other natural mineral fibres.* Geneva, World Health Organization, 1986 (Environmental Health Criteria, No. 53).
2. Peto, J. Dose and time relationships for lung cancer and mesothelioma in relation to smoking and asbestos exposure. *In:* Fischer, M. & Meyer, E., ed. *Zur Beurteilung der Krebsgefahr durch Asbest* [Assessment of the cancer risk of asbestos]. Munich, Medizin Verlag, 1984.
3. Aurand, K. & Kierski, W.-S., ed. *Gesundheitliche Risiken von Asbest. Eine Stellungnahme des Bundesgesundheitsamtes Berlin* [Health risks of asbestos. A position paper of the Federal Health Office, Berlin]. Berlin, Dietrich Reimer Verlag, 1981 (BgA-Berichte, No. 4/81).
4. Schneiderman, M.S. et al. *Assessment of risks posed by exposure to low levels of asbestos in the general environment.* Berlin, Dietrich Reimer Verlag, 1981 (BgA-Bericht, No. 4/81).

5. National Research Council. *Asbestiform fibers: nonoccupational health risks.* Washington, DC, National Academy Press, 1984.
6. Breslow, L. et al. Letter. *Science,* **234**: 923 (1986).
7. *Airborne asbestos health assessment update.* Research Triangle Park, NC, US Environmental Protection Agency, 1985 (Publication EPA-600/8-84-003F).
8. Enterline, P.E. Cancer produced by nonoccupational asbestos exposure in the United States. *Journal of the Air Pollution Control Association,* **33**: 318–322 (1983).
9. Hammond, E.C. et al. Asbestos exposure, cigarette smoking and death rates. *Annals of the New York Academy of Sciences,* **330**: 473–490 (1979).
10. Liddell, F.D.K. Some new and revised risk extrapolations from epidemiological studies on asbestos workers. *In:* Fischer, M. & Meyer, E., ed. *Zur Beurteilung der Krebsgefahr durch Asbest* [Assessment of the cancer risk of asbestos]. Munich, Medizin Verlag, 1984.
11. Liddell, F.D.K. & Hanley, J.A. Relations between asbestos exposure and lung cancer SMRs in occupational cohort studies. *British journal of industrial medicine,* **42**: 389–396 (1985).
12. McDonald, J.C. et al. Dust exposure and mortality in chrysotile mining, 1910–1975. *British journal of industrial medicine,* **37**: 11–24 (1980).
13. Nicholson, W.J. et al. Long-term mortality experience of chrysotile miners and millers in Thetford Mines, Quebec. *Annals of the New York Academy of Sciences,* **330**: 11–21 (1979).
14. Berry, G. & Newhouse, M.L. Mortality of workers manufacturing friction materials using asbestos. *British journal of industrial medicine,* **40**: 1–7 (1983).
15. Henderson, V.L. & Enterline, P.E. Asbestos exposure: factors associated with excess cancer and respiratory disease mortality. *Annals of the New York Academy of Sciences,* **330**: 117–126 (1979).
16. Enterline, P. et al. Mortality in relation to occupational exposure in the asbestos industry. *Journal of occupational medicine,* **14**: 897–903 (1972).
17. Newhouse, M.L. & Berry, G. Patterns of mortality in asbestos factory workers in London. *Annals of the New York Academy of Sciences,* **330**: 53–60 (1979).
18. Weill, H. et al. Influence of dose and fiber type on respiratory malignancy risk in asbestos cement manufacturing. *American review of respiratory diseases,* **120**: 345–354 (1979).
19. Peto, J. Lung cancer mortality in relation to measured dust levels in an asbestos textile factory. *In:* Wagner, J.C., ed. *Biological effects of mineral fibres.* Lyons, International Agency for Research on Cancer, 1980 (IARC Scientific Publications, No. 30).

20. DEMENT, J.M. ET AL. Estimates of dose–response for respiratory cancer among chrysotile asbestos textile workers. *Annals of occupational hygiene,* **26**: 869–887 (1982).
21. FRY, J.S. ET AL. Respiratory cancer in chrysotile production and textile manufacture. *Scandinavian journal of work, environment and health,* **9**: 68–70 (1983).
22. SEIDMAN, H. ET AL. Short-term asbestos work exposure and long-term observation. *Annals of the New York Academy of Sciences,* **330**: 61–67 (1979).
23. SELIKOFF, J.J. ET AL. Mortality experience of insulation workers in the United States and Canada, 1943–1976. *Annals of the New York Academy of Sciences,* **330**: 91–116 (1979).

6.3 Cadmium

Exposure evaluation

It is not possible to carry out a dose–response analysis for cadmium in air solely on the basis of epidemiological data collected in the general population, since the latter is exposed to cadmium mainly via food or tobacco smoking. In addition, the recently reported renal effects in areas of Belgium and the Netherlands polluted by cadmium refer to historical contamination of the environment. Assuming, however, that the only route of exposure is by inhalation, an indirect estimate of the risk of renal dysfunction or lung cancer can be made on the basis of data collected in industrial workers.

Health risk evaluation

Pooled data from seven studies, in which the relationships between the occurrence of tubular proteinuria and cumulative cadmium exposure were examined, show that the prevalence of tubular dysfunction (background level 2.4%) increases sharply at a cumulative exposure of more than 500 $\mu g/m^3$-years (8% at 400 $\mu g/m^3$-years, 50% at 1000 $\mu g/m^3$-years and > 80% at more than 4500 $\mu g/m^3$-years) *(1)*. Some studies suggest that a proportion of workers with cumulative exposures of 100–400 $\mu g/m^3$-years might develop tubular dysfunction (prevalences increasing from 2.4% to 8.8%, at cumulative exposures above 200 $\mu g/m^3$-years). These estimates agree well with that derived from the kinetic model of Kjellström *(2)*, which predicted that the critical concentration of 200 mg/kg in the renal cortex will be reached in 10% of exposed workers after 10 years of exposure to 50 $\mu g/m^3$ and in 1% after 10 years of exposure to 16 $\mu g/m^3$ (cumulative exposures of 500 and 160 $\mu g/m^3$-years, respectively).

With respect to the risk of lung cancer, two risk estimates have been made, one based on the long-term rat bioassay data of Takenaka et al. *(3)* and the other on the epidemiological data of Thun et al. *(4)*. Modelling of these data yielded risk estimates that did not agree. On the basis of the Takenaka data, the unit risk is 9.2×10^{-2} per $\mu g/m^3$; the human data yielded a unit risk of 1.8×10^{-3} per $\mu g/m^3$. In general, the use of human data is more reliable because of species variation in response. Nevertheless, there is evidence from recent studies that this latter unit risk might be substantially overestimated owing to confounding by concomitant exposure to arsenic.

Some uncertainty exists with regard to the thresholds of exposure associated with effects on the kidney. This is primarily due to the limited number of subjects, methodological differences and inaccuracies in exposure data. An overall assessment of the data from industrial workers suggests that, to prevent tubular dysfunction, the 8-hour exposure level for cadmium should not exceed 5 $\mu g/m^3$. This corresponds to a cumulative exposure of 225 $\mu g/m^3$-years. Adopting the lowest estimate of the critical cumulative exposure to airborne cadmium (100 $\mu g/m^3$-years), extrapolation to continuous lifetime exposure results in a permissible concentration of about 300 ng/m^3.

Cadmium in ambient air is transferred to soil by wet or dry deposition and can enter the food chain. However, the rate of transfer from soil to plant depends on numerous factors (type of soil and plant, soil pH, use of fertilizers, meteorology, etc.) and is impossible to predict.

Present average concentrations of cadmium in the renal cortex in the general population in Europe at the age of 40–60 years are in the range 15–40 mg/kg. These values are only 4–12 times lower than the critical levels estimated in cadmium workers for the induction of tubular dysfunction (180 mg/kg) and very close to the critical level of 50 mg/kg estimated by the Cadmibel study in Belgium *(5)*. Any further increase in the dietary intake of cadmium owing to an accumulation of the metal in agricultural soils will further narrow the gap to these critical levels. It is thus imperative to maintain a zero balance for cadmium in agricultural soils by controlling and restricting inputs from fertilizers (including sewage sludge) and atmospheric emissions. Since emissions from industry are currently decreasing, attention must be focused on the emissions from waste incineration, which are likely to increase in the future.

Guidelines

IARC has classified cadmium and cadmium compounds as Group 1 human carcinogens, having concluded that there was sufficient evidence that cadmium can produce lung cancers in humans and animals exposed by inhalation *(6)*. Because of the identified and controversial influence of concomitant exposure to arsenic in the epidemiological study, however, no reliable unit risk can be derived to estimate the excess lifetime risk for lung cancer.

Cadmium, whether absorbed by inhalation or via contaminated food, may give rise to various renal alterations. The lowest estimate of the cumulative exposure to airborne cadmium in industrial workers leading to an increased risk of renal dysfunction (low-molecular-weight proteinuria) or lung cancer is 100 $\mu g/m^3$-years for an 8-hour exposure. Extrapolation to a continuous lifetime exposure gives a value of around 0.3 μg/m^3. Existing levels of cadmium in the air of most urban or industrial areas are around one-fiftieth of this value.

The finding of renal effects in areas contaminated by past emissions of cadmium indicates that the cadmium body burden of the general population in some parts of Europe cannot be further increased without endangering renal function. To prevent any further increase of cadmium in agricultural soils likely to increase the dietary intake of future generations, a guideline of 5 ng/m^3 is established.

References

1. THUN, M. ET AL. Scientific basis for an occupational standard for cadmium. *American journal of industrial medicine,* **20**: 629–642 (1991).
2. KJELLSTRÖM, T. Critical organs, critical concentrations, and whole body dose–response relationships. *In:* Friberg, L. et al., ed. *Cadmium and health: a toxicological and epidemiological appraisal. Vol. 2. Effects and response.* Boca Raton, FL, CRC Press, 1986.
3. TAKENAKA, S. ET AL. Carcinogenicity of cadmium chloride aerosols in Wistar rats. *Journal of the National Cancer Institute,* **70**: 367–371 (1983).
4. THUN, M. ET AL. Mortality among a cohort of U.S. cadmium production workers – an update. *Journal of the National Cancer Institute,* **74**: 325–333 (1985).
5. BUCHET, J.P. ET AL. Renal effects of cadmium body burden of the general population. *Lancet,* **336**: 699–702 (1990).
6. *Beryllium, cadmium, mercury, and exposure in the glass manufacturing industry.* Lyons, International Agency for Research on Cancer, 1993 (IARC Monographs on the Evaluation of Carcinogenic Risks to Humans, Vol. 58).

6.4 Chromium

Exposure evaluation

Chromium is ubiquitous in nature. Available data, generally expressed as total chromium, show a concentration range of 5–200 ng/m^3. There are few valid data on the valency and bioavailability of chromium in the ambient air.

Health risk evaluation

Chromium(III) is recognized as a trace element that is essential to both humans and animals. Chromium(VI) compounds are toxic and carcinogenic, but the various compounds have a wide range of potencies. As the bronchial tree is the major target organ for the carcinogenic effects of chromium(VI) compounds, and cancer primarily occurs following inhalation exposure, uptake in the respiratory organs is of great significance with respect to the cancer hazard and the subsequent risk of cancer in humans. IARC has stated that for chromium and certain chromium compounds there is sufficient evidence of carcinogenicity in humans (Group 1) *(1)*.

A large number of epidemiological studies have been carried out on the association between human exposure to chromates and the occurrence of cancer, particularly lung cancer, but only a few of these include measurements of exposure *(2–8)*. Measurements were made mainly at the time that the epidemiological studies were performed, whereas the carcinogenic effect is caused by exposure dating back 15–30 years. Hence, there is a great need for studies that include historical data on exposure.

Four sets of data for chromate production workers can be used for the quantitative risk assessment of chromium(VI) lifetime exposure *(3, 5–9)*. The average relative risk model is used in the following to estimate the incremental unit risk.

Using the study performed by Hayes et al. on chromium production workers *(3)*, several cohorts were investigated by Braver et al. *(8)* for cumulative exposure to chromium(VI) in terms of µg/m^3-years (cumulative exposure = usual exposure level in µg/m^3 × average duration of exposure). Average lifetime exposures for two cohorts can be calculated from the cumulative exposures of 670 and 3647 µg/m^3-years, as 2 µg/m^3 and 11.4 µg/m^3, respectively (X = µg/m^3 × 8/24 × 240/365 × (No. of years)/70).

The relative risk (RR) for these two cohorts, calculated from observed and expected cases of lung cancer, was 1.75 and 3.04. On the basis of the vital statistics data, the background lifetime probability of death due to lung cancer (P_0) is assumed to be 0.04. The risks (unit risk, UR) associated with a lifetime exposure to 1 μg/m^3 can therefore be calculated to be 1.5×10^{-2} and 7.2×10^{-3}, respectively ($UR = P_0(RR-1)/X$). The arithmetic mean of these two risk estimates is 1.1×10^{-2}.

A risk assessment can also be made on the basis of the study carried out by Langård et al. on ferrochromium plant workers in Norway *(5, 10)*. The chromium concentration to which the workers were exposed is not known, but measurements taken in 1975 showed a geometric mean value of about 530 μg/m^3. Assuming that the content of chromium(VI) in the sample was 19% and previous concentrations were at least as high as in 1975, the ambient concentration would have been about 100 μg/m^3. On the assumption that occupational exposure lasted for about 22 years, the average lifetime exposure can be determined as 6.9 μg/m^3 (X = 100 μg/m^3 × 8/24 × 240/365 × 22/70).

When workers in the same plant who were not exposed to chromium were used as a control population, the relative risk of lung cancer in chromium-exposed workers was calculated to be 8.5. The lifetime unit risk is therefore 4.3×10^{-2}.

Since earlier exposures must have been much higher than the values measured in 1975, the calculated unit risk of 4.3×10^{-2} can only be considered as an upper-bound estimate. The highest relative incidence ever demonstrated in chromate workers in Norway is about 38, at an exposure level for chromium(VI) of about 0.5 mg/m^3 *(6, 7)*. This relative rate is based on the incidence of bronchial cancer of 0.079 in the total Norwegian male population, irrespective of smoking status. If the average exposure duration is about 7 years, the average lifetime daily exposure is calculated to be 11 μg/m^3 (X = 500 μg/m^3 × 8/24 × 240/365 × 7/70). The incremental unit risk was calculated to be 1.3×10^{-1}. This very high lifetime risk may be due to the relatively small working population.

Differences in the epidemiological studies cited may suggest that the different hexavalent chromium compounds have varying degrees of carcinogenic potency.

The estimated lifetime risks based on various epidemiological data sets, in the range of 1.3×10^{-1} to 1.1×10^{-2}, are relatively consistent. As a best estimate, the geometric mean of the risk estimates of 4×10^{-2} may be taken as the incremental unit risk resulting from a lifetime exposure to chromium(VI) at a concentration of 1 μg/m³.

Using some other studies and different risk assessment models, the US Environmental Protection Agency (EPA) estimated the lifetime cancer risk due to exposure to chromium(VI) to be 1.2×10^{-2}. This estimate placed chromium(VI) in the first quartile of the 53 compounds evaluated by the EPA Carcinogen Assessment Group for relative carcinogenic potency *(11)*.

Guidelines

Information on the speciation of chromium in ambient air is essential since, when inhaled, only hexavalent chromium is carcinogenic in humans. The available data are derived from studies among chromium(VI)-exposed workers. When assuming a linear dose–response relationship between exposure to chromium(VI) compounds and lung cancer, no safe level of chromium(VI) can be recommended. At an air concentration of chromium(VI) of 1 μg/m³, the lifetime risk is estimated to be 4×10^{-2}.

It should be noted that chromium concentration in air is often expressed as total chromium and not chromium(VI). The concentrations of chromium(VI) associated with an excess lifetime risk of 1:10 000, 1:100 000 and 1:1 000 000 are 2.5 ng/m³, 0.25 ng/m³ and 0.025 ng/m³, respectively.

References

1. *Chromium, nickel and welding*. Lyons, International Agency for Research on Cancer, 1990 (IARC Monographs on the Evaluation of Carcinogenic Risks to Humans, Vol. 49), pp. 463–474.
2. MACHLE, W. & GREGORIUS, F. Cancer of the respiratory system in the United States chromate-producing industry. *Public health reports*, **63**: 1114–1127 (1948).
3. HAYES, R.B. ET AL. Mortality in chromium chemical production workers: a prospective study. *International journal of epidemiology*, **8**: 365–374 (1979).
4. MANCUSO, T.F. Consideration of chromium as an industrial carcinogen. *In*: Hutchinson, T.C., ed. *Proceedings of the International Conference on Heavy Metals in the Environment, Toronto, 1975*. Toronto, Institute for Environmental Studies, 1975, pp. 343–356.
5. LANGÅRD, S. ET AL. Incidence of cancer among ferrochromium and ferrosilicon workers. *British journal of industrial medicine*, **37**: 114–120 (1980).

6. LANGÅRD, S. & VIGANDER, T. Occurrence of lung cancer in workers producing chromium pigments. *British journal of industrial medicine*, **40**: 71–74 (1983).
7. LANGÅRD, S. & NORSETH, T. A cohort study of bronchial carcinomas in workers producing chromate pigments. *British journal of industrial medicine*, **32**: 62–65 (1975).
8. BRAVER, E.R. ET AL. An analysis of lung cancer risk from exposure to hexavalent chromium. *Teratogenesis, carcinogenesis and mutagenesis*, **5**: 365–378 (1985).
9. HAYES, R. ET AL. Cancer mortality among a cohort of chromium pigment workers. *American journal of industrial medicine*, **16**: 127–133 (1989).
10. LANGÅRD, S. ET AL. Incidence of cancer among ferrochromium and ferrosilicon workers: an extended follow up. *British journal of industrial medicine*, **47**: 14–19 (1990).
11. *Health assessment document for chromium*. Washington, DC, US Environmental Protection Agency, 1984 (Final report EPA-600-8-83-014F).

6.5 Fluoride

Exposure evaluation

Exposure of the general European population to fluoride in its various chemical forms is highly variable. In heavily industrialized urban areas, typical daily inhalation intakes are in the range 10–40 μg/day (0.5–2 μg/m³), and in some cases are as high as 60 μg/day (3 μg/m³). Fluorides are emitted to the atmosphere in both gaseous and particulate forms, but studies typically only report total fluoride content.

The main sources of fluoride intake by humans are food and water. Except for occupational exposure, exposure to fluoride by inhalation is negligible.

Regarding occupational exposure, the daily amount of fluoride inhaled, assuming a total respiratory rate of 10 m³ during a working day, could be 10–25 mg when the air concentration is at the most frequent exposure limits of 1–2.5 mg/m³.

Health risk evaluation

The most important long-term adverse effect of fluorides on human populations is endemic skeletal fluorosis. The beneficial effect is prevention of caries, as a result both of fluoride incorporation into developing teeth and post-eruptive exposure of enamel to adequate levels of fluoride. It is therefore of crucial importance to gather information on fluoride sources in the diet, especially water, the etiology of early skeletal fluorosis as related to bone mineralization, and dose–response relationships *(1)*.

The earliest reports of skeletal fluorosis appeared from industries where exposure of workers to 100–500 μg/m³ per 8-hour day for more than 4 years led to severe skeletal changes. Skeletal fluorosis has also been diagnosed in persons living in areas with excessive fluoride in soil, water, dust or plants *(1)*.

In one study, bronchial hyperreactivity was the main health effect at a mean fluoride concentration of 0.56 mg/m³ and a mean particulate fluoride concentration of 0.15 mg/m³ *(2)*. In a longitudinal study performed on 523 aluminium potroom workers, total fluoride was the most important risk factor among the exposure variables. In this study, the risk of developing asthmatic symptoms such as dyspnoea and wheezing was 3.4 and 5.2 times higher in the medium- and high-exposure groups, respectively, than in the low-exposure group. Exposure to other pollutants was limited and did not appear to confound the results *(3)*.

Children living in the vicinity of a phosphate processing facility who were exposed to concentrations of about 100–500 μg/m³ exhibited an impairment of respiratory function. It is not known, however, whether the concentrations were gaseous or total fluoride. In another study, no effects on respiratory function were observed at gaseous fluoride levels of up to 16 μg/m³.

There is no evidence that atmospheric deposition of fluorides results in significant exposure through other routes, such as through contamination of soil and consequently groundwater.

Guidelines

For exposure of the general population to fluoride, reference exposure levels have been derived by applying a "benchmark dose" approach to a variety of animal and human exposure studies. The 1-hour reference exposure level to protect against any respiratory irritation is about 0.6 mg/m³, and the level to protect against severe irritation from a once-in-a-lifetime release is about 1.6 mg/m³ *(4)*.

Data from various sources indicate that prolonged exposure of humans (workers and children) to fluoride concentrations of 0.1–0.5 mg/m³ leads to impairment of pulmonary function and skeletal fluorosis. No effects have been found at levels of up to 16 μg/m³ gaseous fluoride. However, the available information does not permit the derivation of an air quality guideline value for fluoride(s).

Skeletal fluorosis is associated with a systemic uptake exceeding 5 mg/day in a relatively sensitive section of the general population. Systemic uptake from food and fluoridated water is about 3 mg/day. It is highly unlikely that ambient air concentrations of fluorides could pose any material risk of fluorosis.

It has been recognized that fluoride levels in ambient air should be less than 1 μg/m³ to prevent effects on livestock and plants. These concentrations will also sufficiently protect human health.

References

1. *Fluorine and fluorides*. Geneva, World Health Organization, 1984 (Environmental Health Criteria, No. 36).

2. SARIC, M. ET AL. The role of atopy in potroom workers' asthma. *American journal of industrial medicine,* **9**: 239–242 (1986).
3. KONGERUD, J. & SAMUELSEN, S.O. A longitudinal study of respiratory symptoms in aluminum potroom workers. *American review of respiratory diseases,* **144**: 10–16 (1991).
4. ALEXEEF, G.V. ET AL. Estimation of potential health effects from acute exposure to hydrogen fluoride using a "benchmark dose" approach. *Risk analysis,* **13**: 63–69 (1993).

6.6 Hydrogen sulfide

Exposure evaluation

Typical symptoms and signs of hydrogen sulfide intoxication are most often caused by relatively high concentrations in occupational exposures. There are many occupations where there is a potential risk of hydrogen sulfide intoxication and, according to the US National Institute for Occupational Safety and Health *(1)*, in the United States alone approximately 125 000 employees are potentially exposed to hydrogen sulfide. Low-level concentrations can occur more or less continuously in certain industries, such as in viscose rayon and pulp production, at oil refineries and in geothermal energy installations.

In geothermal areas there is a risk of exposure to hydrogen sulfide for the general population *(2)*. The biodegradation of industrial wastes has been reported to cause ill effects in the general population *(2)*. An accidental release of hydrogen sulfide into the air surrounding industrial facilities can cause very severe effects, as at Poza Rica, Mexico, where 320 people were hospitalized and 22 died *(2)*. The occurrence of low-level concentrations of hydrogen sulfide around certain industrial installations is a well known fact.

Health risk evaluation

The first noticeable effect of hydrogen sulfide at low concentrations is its unpleasant odour. Conjunctival irritation is the next subjective symptom and can cause so-called "gas eye" at hydrogen sulfide concentrations of 70–140 mg/m^3. Table 16 shows the established dose–effect relationships for hydrogen sulfide.

The hazards caused by high concentrations of hydrogen sulfide are relatively well known, but information on human exposure to very low concentrations is scanty. Workers exposed to hydrogen sulfide concentrations of less than 30 mg/m^3 are reported to have rather diffuse neurological and mental symptoms *(4)* and to show no statistically significant differences when compared with a control group. On the other hand, changes in haem synthesis have been reported at hydrogen sulfide concentrations of less than 7.8 mg/m^3 (1.5–3 mg/m^3 average) *(5)*. It is not known whether the inhibition is caused by the low concentrations or by the cumulative effects of occasional peak concentrations. Most probably, at concentrations below 1.5 mg/m^3 (1 ppm), even with exposure for longer periods, there are very few detectable health hazards in the toxicological sense. The malodorous

Table 16. Hydrogen sulfide: established dose–effect relationships

Hydrogen sulfide concentration		Effect	Reference
mg/m³	ppm		
1400–2800	1000–2000	Immediate collapse with paralysis of respiration	*(2)*
750–1400	530–1000	Strong central nervous system stimulation, hyperpnoea followed by respiratory arrest	*(2)*
450–750	320–530	Pulmonary oedema with risk of death	*(2)*
210–350	150–250	Loss of olfactory sense	*(3)*
70–140	50–100	Serious eye damage	*(3)*
15–30	10–20	Threshold for eye irritation	*(3)*

property of hydrogen sulfide is a source of annoyance for a large proportion of the general population at concentrations below 1.5 mg/m³, but from the existing data it cannot be concluded whether any health effects result. The need for epidemiological studies on possible effects of long-term, low-level hydrogen sulfide exposure is obvious. A satisfactory biological exposure indicator is also needed.

Guidelines

The LOAEL of hydrogen sulfide is 15 mg/m³, when eye irritation is caused. In view of the steep rise in the dose–effect curve implied by reports of serious eye damage at 70 mg/m³, an uncertainty factor of 100 is recommended, leading to a guideline value of 0.15 mg/m³ with an averaging time of 24 hours. A single report of changes in haem synthesis at a hydrogen sulfide concentration of 1.5 mg/m³ should be borne in mind.

In order to avoid substantial complaints about odour annoyance among the exposed population, hydrogen sulfide concentrations should not be allowed to exceed 7 μg/m³, with a 30-minute averaging period.

When setting concentration limits in ambient air, it should be remembered that in many places hydrogen sulfide is emitted from natural sources.

References

1. *Occupational exposure to hydrogen sulfide.* Cincinnati, OH, US Department of Health, Education, and Welfare, 1977 (DHEW Publication (NIOSH) No. 77-158).
2. *Hydrogen sulfide.* Geneva, World Health Organization, 1981 (Environmental Health Criteria, No. 19).
3. SAVOLAINEN, H. Nordiska expertgruppen för gränsvärdesdokumentation. 40. Dihydrogensulfid [Nordic expert group for TLV evaluation. 40. Hydrogen sulfide]. *Arbeta och hälsa,* **31**: 1–27 (1982).
4. KANGAS, J. ET AL. Exposure to hydrogen sulfide, mercaptans and sulfur dioxide in pulp industry. *American Industrial Hygiene Association journal,* **45**: 787–790 (1984).
5. TENHUNEN, R. ET AL. Changes in haem synthesis associated with occupational exposure to organic and inorganic sulphides. *Clinical science,* **64**: 187–191 (1983).

6.7 Lead

Exposure evaluation

Average air lead levels are usually below 0.15 μg/m³ at nonurban sites. Urban air lead levels are typically between 0.15 and 0.5 μg/m³ in most European cities *(1–3)*. Additional routes of exposure must not be neglected, such as lead in dust, a cause of special concern for children.

The relationship between air lead exposure and blood lead has been shown to exhibit downward curvilinearity if the range of exposures is sufficiently large. At lower levels of exposure, the deviation from linearity is negligible and linear models of the relationship between intake and blood lead are satisfactory approximations.

The level of lead in blood is the best available indicator of current and recent past environmental exposure, and may also be a reasonably good indicator of lead body burden with stable exposures. Biological effects of lead will, therefore, be related to blood lead as an indicator of internal exposure.

Health risk evaluation

Table 17 summarizes LOAELs for haematological and neurological effects in adults. Cognitive effects in lead workers have not been observed at blood lead levels below 400 μg/l *(4, 5)*. Reductions in nerve conduction velocity were found in lead workers at blood levels as low as 300 μg/l *(6–8)*. Elevation of free erythrocyte protoporphyrin has been observed at blood levels of 200–300 μg/l. Delta-aminolaevulinic acid dehydrase (ALAD) inhibition is likely to occur at blood levels of about 100 μg/l *(9)*. Because of its uncertain biological significance relative to the functional reserve capacity of the haem biosynthetic system, ALAD inhibition is not treated as an adverse effect here.

Table 18 summarizes LOAELs for haematological, endocrinological and neurobehavioural endpoints in children. Reduced haemoglobin levels have been found at concentrations in blood of around 400 μg/l. Haematocrit values below 35% have not been reported at blood levels below 200 μg/l *(10)*; this is also true for several enzyme systems, which may be of clinical significance.

Central nervous system effects, as assessed by neurobehavioural endpoints, appear to occur at levels below 200 μg/l. Consistent effects have been

Table 17. Summary of LOAELs for lead-induced health effects in adults

LOAEL at given blood lead level (μg/l)	Haem synthesis, haematological and other effects	Effects on the nervous system
1000–1200		Encephalopathic signs and symptoms
800	Frank anaemia	
500	Reduced haemoglobin production	Overt subencephalopathic neurological symptoms, cognition impairment
400	Increased urinary ALA and elevated coproporphyrin	
300		Peripheral nerve dysfunction (slowed nerve conduction velocities)
200–300	Erythrocyte protoporphyrin elevation in males	
150–200	Erythrocyte protoporphyrin elevation in females	

Table 18. Summary of LOAELs for lead-induced health effects in children

LOAEL at given blood lead level (μg/l)	Haem synthesis, haematological and other effects	Effects on the nervous system
800–1000		Encephalopathic signs and symptoms
700	Frank anaemia	
400	Increased urinary delta-aminolaevulinic acid and elevated coproporphyrin	
250–300	Reduced haemoglobin synthesis	
150–200	Erythrocyte protoporphyrin elevation	
100–150	Vitamin D3 reduction	Cognitive impairment
100	ALAD inhibition	Hearing impairment

reported for global measures of cognitive functioning, such as the psychometric IQ, to be associated with blood lead levels of 100–150 μg/l *(11, 12)*. Some epidemiological studies have indicated effects at blood lead levels below 100 μg/l. Existing animal studies do provide qualitative support for the claim of lead as the causative agent *(12)*.

Guidelines

Guidelines for lead in air will be based on the concentration of lead in blood. Critical effects to be considered in the adult organism include elevation of free erythrocyte protoporphyrin, whereas for children cognitive deficit, hearing impairment and disturbed vitamin D metabolism *(13, 14)* are taken as the decisive effects. All of these effects are considered adverse. A critical level of lead in blood of 100 μg/l is proposed. It should be stressed that all of these values are based on population studies yielding group averages, which apply to the individual child only in a probabilistic manner. Although some lead salts have been found to be carcinogenic in animals, the evidence for a carcinogenic potential in humans is inadequate and will, therefore, not be considered here.

For the derivation of a guideline value, the following arguments have been considered.

- Currently measured "baseline" blood lead levels of minimal anthropogenic origin are probably in the range10–30 μg/l.

- Various international expert groups have determined that the earliest adverse effects of lead in populations of young children begin at 100–150 μg/l. Although it cannot be excluded that population effects may occur below this range, it is assumed to be prudent to derive a guideline value based on the lowest value in this range (100 μg/l).

- It can be assumed that inhalation of airborne lead is a significant route of exposure for adults (including pregnant women) but is of less significance for young children, for whom other pathways of exposure such as ingested lead are generally more important.

- It appears that 1 μg lead per m^3 air directly contributes approximately 19 μg lead per litre blood in children and about 16 μg per litre blood in adults, although it is accepted that the relative contribution from air is less significant in children than in adults. These values are approximations, recognizing that the relationships are curvilinear in nature and will apply principally at lower blood lead levels.

- It must be taken into account that, in typical situations, an increase of lead in air also contributes to increased lead uptake by indirect environmental pathways. To correct for uptake by other routes as well, it is assumed that 1 μg lead per m^3 air would contribute to 50 μg lead per litre blood.

- It is recommended that efforts be made to ensure that at least 98% of an exposed population, including preschool children, have blood lead levels that do not exceed 100 μg/l. In this case, the median blood lead level would not exceed 54 μg/l. On this basis, the annual average lead level in air should not exceed 0.5 μg/m^3. This proposal is based on the assumption that the upper limit of nonanthropogenic blood is 30 μg/l. These estimates are assumed to protect adults also.

- To prevent further increases of lead in soils and consequent increases in the exposure of future generations, air lead levels should be kept as low as possible.

Since both direct and indirect exposure of young children to lead in air occurs, the air guidelines for lead should be accompanied by other preventive measures. These should specifically take the form of monitoring the lead content of dust and soils arising from lead fallout. The normal hand-to-mouth behaviour of children with regard to dust and soil defines these media as potentially serious sources of exposure. A specific monitoring value is not recommended. Some data indicate that lead fallout in excess of 250 μg/m^2 per day will increase blood lead levels.

References

1. DELUMYEA, R. & KALIVRETENOS, A. Elemental carbon and lead content of fine particles from American and French cities of comparable size and industry, 1985. *Atmospheric environment*, **21**: 1643–1647 (1987).
2. MINISTERIUM FÜR UMWELT, RAUMORDNUNG UND LANDWIRTSCHAFT DES LANDES NW. *Luftreinhaltung in Nordrhein-Westfalen. Eine Erfolgsbilanz der Luftreinhalteplanung 1975–1988*. Bonn, Bonner Universitätsdruckerei, 1989.
3. DUCOFFRE, G. ET AL. Lowering time trend of blood lead levels in Belgium since 1978. *Environmental research*, **51**: 25–34 (1990).
4. STOLLERY, B.T. ET AL. Cognitive functioning in lead workers. *British journal of industrial medicine*, **46**: 698–707 (1989).
5. STOLLERY, B.T. ET AL. Short term prospective study of cognitive functioning in lead workers. *British journal of industrial medicine*, **48**: 739–749 (1991).

6. SEPPÄLÄINEN, A.M. ET AL. Subclinical neuropathy at "safe" levels of lead exposure. *Archives of environmental health*, **30**: 180–183 (1975).
7. SEPPÄLÄINEN, A.M. & HERNBERG, S. A follow-up study of nerve conduction velocities in lead-exposed workers. *Neurobehavioral toxicology and teratology*, **4**: 721–723 (1982).
8. DAVIS, J.M. & SVENDSGAARD, D.J. Nerve conduction velocity and lead: a critical review and meta-analysis. *In*: Johnson, B.L., ed. *Advances in neurobehavioral toxicology*. Chelsea, Lewis Publishers, 1990, pp. 353–376.
9. HERNBERG, S. & NIKKANEN, J. Enzyme inhibition by lead under normal urban conditions. *Lancet*, **1**: 63–64 (1970).
10. SCHWARTZ, J. ET AL. Lead-induced anemia: dose–response relationships and evidence for a threshold. *American journal of public health*, **80**: 165–168 (1990).
11. SCHWARTZ, J. Low-level lead exposure and children's IQ: a meta-analysis and search for a threshold. *Environmental research*, **65**: 42–55 (1994).
12. *Inorganic lead*. Geneva, World Health Organization, 1995 (Environmental Health Criteria, No. 165).
13. MAHAFFEY, K.R. ET AL. Association between age, blood lead concentration, and serum 1,25-dihydroxycholealciferol levels in children. *American journal of clinical nutrition*, **35**: 1327–1331 (1982).
14. ROSEN, J.F. ET AL. Reduction in 1,25-dihydroxyvitamin D in children with increased lead absorption. *New England journal of medicine*, **302**: 1128–1131 (1980).

6.8 Manganese

Exposure evaluation

In urban and rural areas without significant manganese pollution, annual averages are mainly in the range of 0.01–0.07 $\mu g/m^3$; near foundries the level can rise to an annual average of 0.2–0.3 $\mu g/m^3$ and, where ferro- and silico-manganese industries are present, to more than 0.5 $\mu g/m^3$, with individual 24-hour concentrations sometimes exceeding 10 $\mu g/m^3$ *(1, 2)*.

Health risk evaluation

The toxicity of manganese varies according to the route of exposure. By ingestion, manganese has relatively low toxicity at typical exposure levels and is considered a nutritionally essential trace element. By inhalation, however, manganese has been known since the early nineteenth century to be toxic to workers. Manganism is characterized by various psychiatric and movement disorders, with some general resemblance to Parkinson's disease in terms of difficulties in the fine control of some movements, lack of facial expression, and involvement of underlying neuroanatomical (extrapyramidal) and neurochemical (dopaminergic) systems *(3–5)*. Respiratory effects such as pneumonitis and pneumonia and reproductive dysfunction such as reduced libido are also frequently reported features of occupational manganese intoxication. The available evidence is inadequate to determine whether or not manganese is carcinogenic; some reports suggest that it may even be protective against cancer. Based on this mixed but insufficient evidence, the US Environmental Protection Agency has concluded that manganese is not classifiable as to human carcinogenicity *(6)*. IARC has not evaluated manganese *(7)*.

Several epidemiological studies of workers have provided consistent evidence of neurotoxicity associated with low-level manganese exposure. Sufficient information was available to develop a benchmark dose using the study by Roels et al. *(3)*, thereby obviating the need to account for a LOAEL to NOAEL extrapolation. With regard to exposure, both lifetime integrated respirable dust concentrations as well as current respirable dust concentrations were considered. Correlation between effects and exposure was strongest for eye–hand coordination with current concentration of respirable dust. From the data of Roels et al. *(3)*, lower 95% confidence limits of the best concentration estimate giving respectively a 10% effect ($BMDL_{10}$) of 74 $\mu g/m^3$ and a 5% effect ($BMDL_5$) of 30 $\mu g/m^3$ were calculated *(8)*. Taking a conservative approach, the lower 95% confidence limit of the $BMDL_5$ values was chosen as representative of the NOAEL.

$BMDL_5$ values for the other exposure measures (time-integrated and average concentration of respirable dust) are not substantially different *(5)*.

In evaluating the potential health risks associated with inhalation exposure to manganese, various uncertainties must be taken into consideration. Virtually all of the human health evidence is based on healthy, adult male workers; other, possibly more sensitive populations have not been adequately investigated. Also, the potential reproductive and developmental toxicity of inhaled manganese has not been fully investigated.

Guidelines

Based on neurotoxic effects observed in occupationally exposed workers and using the benchmark approach, an estimated NOAEL (the lower 95% confidence limit of the $BMDL_5$) of 30 $\mu g/m^3$ was obtained. A guideline value for manganese of 0.15 $\mu g/m^3$ was derived by dividing by a factor of 4.2 to adjust for continuous exposure and an uncertainty factor of 50 (10 for interindividual variation and 5 for developmental effects in younger children). This latter factor was chosen by analogy with lead where neurobehavioural effects were found in younger children at blood lead levels five times lower than in adults and supported by evidence from studies of experimental animals. The adjustment for continuous exposure was considered sufficient to account for long-term exposure based on knowledge of the half-time of manganese in the brain. The guideline value should be applied as an annual average.

References

1. *Reevaluation of inhalation health risks associated with methylcyclopentadienyl manganese tricarbonyl (MMT) in gasoline*. Washington, DC, US Environmental Protection Agency, 1994.
2. Pace, T. G. & Frank, N. H. *Procedures for estimating probability of nonattainment of a PM_{10} NAAQS using total suspended particulate or inhalable particulate data*. Research Triangle Park, NC, US Environmental Protection Agency, 1983.
3. Roels, H. A. et al. Assessment of the permissible exposure level to manganese in workers exposed to manganese dioxide dust. *British journal of industrial medicine*, **49**: 25–34 (1992).
4. Iregren, A. Psychological test performance in foundry workers exposed to low levels of manganese. *Neurotoxicology and teratology*, **12**: 673–675 (1990).
5. Mergler, D. et al. Nervous system dysfunction among workers with long-term exposure to manganese. *Environmental research*, **64**: 151–180 (1994).

6. INTEGRATED RISK INFORMATION SYSTEM (IRIS). *Carcinogenicity assessment for lifetime exposure to manganese* (http://www.epa.gov/ngispgm3/iris/subst/0373.htm#II). Cincinnati, OH, US Environmental Protection Agency (accessed 25 May 1988).
7. BOFFETTA, P. Carcinogenicity of trace elements with reference to evaluations made by the International Agency for Research on Cancer. *Scandinavian journal of work, environment and health*, **19** (Suppl. 1): 67–70 (1993).
8. SLOB, W. ET AL. *Review of the proposed WHO air quality guideline for manganese*. Bilthoven, National Institute of Public Health and Environmental protection (RIVM), 1996 (Report No. 6135100001).

6.9 Mercury

Exposure evaluation

In areas remote from industry, atmospheric levels of mercury are about 2–4 ng/m^3, and in urban areas about 10 ng/m^3. This means that the daily amount absorbed into the bloodstream from the atmosphere as a result of respiratory exposure is about 32–64 ng in remote areas, and about 160 ng in urban areas. However, this exposure to mercury from outdoor air is marginal compared to exposure from dental amalgams, given that the estimated average daily absorption of mercury vapour from dental fillings varies between 3000 and 17 000 ng.

Health risk evaluation

Sensitive population groups

With regard to exposure to mercury vapour, sensitive population groups have not been conclusively identified from epidemiological, clinical or experimental studies. Nevertheless, the genetic expression of the enzyme catalase, which catalyses the oxidation of mercury vapour to divalent mercuric ion, varies throughout populations. Swiss and Swedish studies have revealed a gene frequency of the order of 0.006 for this trait *(1, 2)*. Thus 30–40 per million of the population are almost completely lacking catalase activity (homozygotes) and 1.2% are heterozygotes with a 60% reduction in catalase activity. Information is lacking on the degree to which other enzymes in the blood are able to take over the oxidation.

Effects on the kidney of inorganic mercury and phenylmercury are believed to occur first in a subgroup of individuals whose susceptibility may be genetically determined, although the proportion of this subgroup in the general population is unknown. Virtually nothing is known about the relative sensitivity at different stages of the life cycle to mercury vapour or inorganic cationic compounds, except that the developing rat kidney is less sensitive than the mature tissue to inorganic mercury *(3)*.

The prenatal stage appears to be the period of life when sensitivity to methylmercury is at its greatest; neuromotor effects in exposed Iraqi populations indicated that sensitivity at this time is at least three times greater than that in adults *(4)*.

Mercury vapour

Time-weighted air concentrations are the usual means of assessing human exposure. Reported air values depend on the type of sampling. Static sampling generally gives lower values than personal sampling. In order to convert the air concentrations quoted in Table 19 to equivalent concentrations in ambient air, two factors have to be taken into account. First, the air concentrations listed in Table 19 were measured in the working environment using static samplers. The conversion factor may vary, depending on exposure conditions. The values shown should be increased by a factor of 3 to correspond to the true air concentrations inhaled by the workers as determined by personal samplers. Second, the total amount of air inhaled at the workplace per week is assumed to be 50 m^3 (10 m^3/day × 5 days) whereas the amount of ambient air inhaled per week would be 140 m^3 (20 m^3/day × 7 days). Thus the volume of ambient air inhaled per week is approximately three times the volume inhaled at the workplace. Thus, to convert the workplace air concentrations quoted in Table 19 to equivalent ambient air concentrations, they should first be multiplied by 3 to convert to actual concentrations in the workplace, and divided by 3 to correct for the greater amount of ambient air inhaled per week by the average adult. It follows that the mercury vapour concentrations quoted in Table 19 are approximately equivalent to ambient air concentrations.

Table 19. Concentrations of total mercury in air and urine at which effects are observed at a low frequency in workers subjected to long-term exposure to mercury vapour

Observed effect [a]	Mercury level		Reference
	Air [b] (μg/m^3)	Urine (μg/litre)	
Objective tremor	30	100	*(5)*
Renal tubular effects; changes in plasma enzymes	15 [c]	50	*(6)*
Nonspecific symptoms	10–30	25–100	*(5)*

[a] These effects occur with low frequency in occupationally exposed groups. Other effects have been reported, but air and urine levels are not available.

[b] The air concentrations measured by static air samplers are taken as a time-weighted average, assuming 40 hours per week for long-term exposure (at least five biological half-times, equivalent to 250 days).

[c] Calculated from the urine concentration, assuming that a mercury concentration in air of 100 μg/m^3 measured by static samplers is equivalent to a mercury concentration of 300 μg/litre in the urine.

Since these figures are based on observations in humans, an uncertainty factor of 10 would seem appropriate. However, the LOAELs in Table 19 are rough estimates of air concentrations at which effects occur at a "low frequency". Because it seems unlikely that such effects would occur in occupationally exposed workers at air concentrations as low as one half of those given in Table 19, it seems appropriate to use an uncertainty factor of 20. Thus, the estimated guideline for mercury concentration in air would be 1 μg/m³.

Inorganic compounds

Cationic forms of inorganic mercury are retained in the lungs about half as efficiently as inhaled mercury vapour (40% versus 80% retained); thus the estimated guideline providing adequate protection against renal tubular effects would be twice as high as that for mercury vapour.

Methylmercury compounds

It does not seem appropriate to set air quality guidelines for methylmercury compounds. Inhalation of this form of mercury, if it is present in the atmosphere, would make a negligible contribution to total human intake. Nevertheless, mercury in the atmosphere may ultimately be converted to methylmercury following deposition on soils or sediments in natural bodies of water, leading to an accumulation of that form of mercury in aquatic food chains. In this situation, guidelines for food intake would be appropriate, such as those recommended by the Joint FAO/WHO Expert Committee on Food Additives.

Guidelines

It is necessary to take into account the different forms of mercury in the atmosphere and the intake of these forms of mercury from other media. The atmosphere and dental amalgam are the sole sources of exposure to mercury vapour, whereas the diet is the dominant source of methylmercury compounds.

Current levels of mercury in outdoor air, except for regional "hot spots", are typically in the order of 0.005–0.010 μg/m³ and thus are marginal compared to exposure from dental amalgam. The exposure to mercury from outdoor air at these air levels is not expected to have direct effects on human health.

The predominant species of mercury present in air, Hg^0, is neither mutagenic nor carcinogenic. Exposure to airborne methylmercury is 2–3 orders of magnitude below the food-related daily intake and will, in this context, be regarded as insignificant. It is thus only possible to derive a numerical guideline for inhalation of inorganic mercury, by including mercury vapour and divalent mercury.

The LOAELs for mercury vapour are around 15–30 μg/m^3. Applying an uncertainty factor of 20 (10 for uncertainty due to variable sensitivities in higher risk populations and, on the basis of dose–response information, a factor of 2 to extrapolate from a LOAEL to a likely NOAEL), a guideline for inorganic mercury vapour of 1 μg/m^3 as an annual average has been established. Since cationic inorganic mercury is retained only half as much as the vapour, the guideline also protects against mild renal effects caused by cationic inorganic mercury. Present knowledge suggests, however, that effects on the immune system at lower exposures cannot be excluded.

An increase in ambient air levels of mercury will result in an increase in deposition in natural bodies of water, possibly leading to elevated concentrations of methylmercury in freshwater fish. Such a contingency might have an important bearing on acceptable levels of mercury in the atmosphere. Unfortunately, the limited knowledge of the global cycle and of the methylation and bioaccumulation pathways in the aquatic food chain does not allow any quantitative estimates of risks from these post-depositional processes. Therefore, an ambient air quality guideline value that would fully prevent the potential for adverse health impacts of post-depositional methylmercury formation cannot be proposed. To prevent possible health effects in the near future, however, ambient air levels of mercury should be kept as low as possible.

References

1. AEBI, H. Investigation of inherited enzyme deficiencies with special reference to acatalasia. *In:* Crow, J.F. & Neel, J.V., ed. *Proceedings from 3rd International Congress of Human Genetics, Chicago 1966.* Baltimore, MD, Johns Hopkins, 1967, p.189.
2. PAUL, K.G. & ENGSTEDT, L.M. Normal and abnormal catalase activity in adults. *Scandinavian journal of clinical laboratory investigation,* **10**: 26 (1958).
3. DASTON, G.P. ET AL. Toxicity of mercuric chloride to the developing rat kidney. I. Post-natal ontogeny of renal sensitivity. *Toxicology and applied pharmacology,* 71: 24–41(1983).
4. AL-SHAHRISTANI, H. & SHIHAB, K.M. Variation of biological half-life of methylmercury in man. *Archives of environmental health,* **18**: 342–344 (1974).

5. *Inorganic mercury*. Geneva, World Health Organization, 1991 (Environmental Health Criteria, No. 118).
6. CÁRDENAS, A. ET AL. Markers of early renal changes induced by industrial pollutants. I. Application to workers exposed to mercury vapour. *British journal of industrial medicine*, **50**: 17–27 (1993).

6.10 Nickel

Exposure evaluation

Nickel is present throughout nature and is released into air and water both from natural sources and as a result of human activity.

In nonsmokers, about 99% of the estimated daily nickel absorption stems from food and water; for smokers the figure is about 75%. Nickel levels in the ambient air are in the range 1–10 ng/m^3 in urban areas, although much higher levels (110–180 ng/m^3) have been recorded in heavily industrialized areas and larger cities. There is, however, limited information on the species of nickel in ambient air.

Consumer products made from nickel alloys and nickel-plated items lead to cutaneous contact exposure.

Exposure to nickel levels of 10–100 mg/m^3 have been recorded for occupational groups, with documented increased cancer risk. Exposure levels in the refining industry are currently usually less than 1–2 mg/m^3, often less than 0.5 mg/m^3. Experimental and epidemiological data indicate that the nickel species in question is important for risk estimation.

Health risk evaluation

Allergic skin reactions are the most common health effect of nickel, affecting about 2% of the male and 11% of the female population. Nickel content in consumer products and possibly in food and water are critical for the dermatological effect. The respiratory tract is also a target organ for allergic manifestations of occupational nickel exposure.

Work-related exposure in the nickel-refining industry has been documented to cause an increased risk of lung and nasal cancers. Inhalation of a mixture of oxidic, sulfidic and soluble nickel compounds at concentrations higher than 0.5 mg/m^3, often considerably higher, for many years has been reported *(1)*.

Nickel has a strong and prevalent allergenic potency. There is no evidence that airborne nickel causes allergic reactions in the general population, although this reaction is well documented in the working environment. The key criterion for assessing the risk of nickel exposure is its carcinogenic potential.

In general, nickel compounds give negative results in short-term bacterial mutagenicity tests because of limited uptake. Nevertheless, they show a wide range of transformation potencies in mammalian cell assays, depending mainly on their bioavailability.

Both green nickel oxide and the subsulfide have caused tumours in animal inhalation studies. In addition, nickel monoxide (not further specified) and an alloy with 66.5% nickel and 12.5% chromium caused tumours following tracheal instillation. A corresponding instillation with an alloy of 26.8% nickel and 16.2% chromium had no such effect, indicating that it was nickel and not chromium that caused the tumours. Injection-site tumours in a number of organs are found with many particulate nickel compounds. The tumorigenic potency varies with chemical composition, solubility and particle surface properties *(2, 3)*.

Epidemiological evidence from the nickel-refining industry indicates that sulfidic, oxidic and soluble nickel compounds are all carcinogenic. Exposure to metallic nickel has not been demonstrated to cause cancer in workers.

Several theories have been suggested for the mechanisms of nickel tumorigenesis. All of these assume that the nickel ion is the ultimate active agent. On the basis of the underlying concept that all nickel compounds can generate nickel ions that are transported to critical sites in target cells, IARC has classified nickel compounds as carcinogenic to humans (Group 1) and metallic nickel as possibly carcinogenic to humans (Group 2B) *(4)*.

On the basis of one inhalation study *(5)*, the US Environmental Protection Agency (EPA) classified nickel subsulfide as a class A carcinogen and estimated the maximum likelihood incremental unit risk to be $1.8–4.1 \times 10^{-3}$ *(6)*. This study, however, involves only exposure to nickel subsulfide. It is not known whether this compound is present in ambient air, but since it is probably one of the most potent nickel compounds, this risk estimate may represent an upper limit, if accepted. WHO estimated an incremental unit risk of 4×10^{-4} per $\mu g/m^3$ calculated from epidemiological results *(7)*.

On the basis of epidemiological studies, EPA classified nickel dust as a class A carcinogen and estimated the lifetime cancer risk from exposure to nickel dust to be 2.4×10^{-4}. This estimate placed nickel in the third quartile of the 55 substances evaluated by the EPA Carcinogen Assessment Group with regard to their relative carcinogenic potency *(8)*. Assuming a content of 50% of nickel subsulfide in total dust, a unit risk of 4.8×10^{-4} was estimated for this compound.

An estimate of unit risk can be given on the basis of the report of lung cancer in workers first employed between 1968 and 1972 and followed through to 1987 in Norway *(9, 10)*. Using the estimated risk of 1.9 for this group and an exposure of 2.5 mg/m^3, a lifetime exposure of 155 μg/m^3 and a unit risk of 3.8×10^{-4} per μg/m^3 can be calculated.

Guidelines

Even if the dermatological effects of nickel are the most common, such effects are not considered to be critically linked to ambient air levels.

Nickel compounds are human carcinogens by inhalation exposure. The present data are derived from studies in occupationally exposed human populations. Assuming a linear dose–response, no safe level for nickel compounds can be recommended.

On the basis of the most recent information of exposure and risk estimated in industrial populations, an incremental risk of 3.8×10^{-4} can be given for a concentration of nickel in air of 1 μg/m^3. The concentrations corresponding to an excess lifetime risk of 1:10 000, 1:100 000 and 1: 1 000 000 are about 250, 25 and 2.5 ng/m^3, respectively.

References

1. Report of the International Committee on Nickel Carcinogenesis in Man. *Scandinavian journal of work, environment and health*, **16**: 1–82 (1990).
2. SUNDERMAN, F.W., JR. Search for molecular mechanisms in the genotoxicity of nickel. *Scandinavian journal of work, environment and health*, **19**: 75–80 (1993).
3. COSTA, M. ET AL. Molecular mechanisms of nickel carcinogenesis. *Science of the total environment*, **148**: 191–200 (1994).
4. Nickel and nickel compounds. *In: Chromium, nickel and welding*. Lyons, International Agency for Research on Cancer, 1990, pp. 257–445 (IARC Monographs on the Evaluation of Carcinogenic Risks to Humans, Vol. 49).
5. OTTOLENGHI, A.D. ET AL. Inhalation studies of nickel sulfide in pulmonary carcinogenesis of rats. *Journal of the National Cancer Institute*, **54**: 1165–1172 (1974).
6. INTEGRATED RISK INFORMATION SYSTEM (IRIS). *Reference concentration (RfC) for inhalation exposure for nickel subsulfide* (http://www.epa.gov/ngispgm3/iris/subst/0273.htm).Cincinnati, OH, US Environmental Protection Agency (accessed 1 April 1987).
7. Nickel. *In: Air Quality Guidelines for Europe*. Copenhagen, WHO Regional Office for Europe, 1987 (WHO Regional Publications, European Series, No. 23), pp. 285–296.

8. INTEGRATED RISK INFORMATION SYSTEM (IRIS). *Reference concentration (RfC) for inhalation exposure for nickel refinery dust* (http://www.epa.gov/ngispgm3/iris/subst/0272.htm). Cincinnati, OH, US Environmental Protection Agency (accessed 1 April 1987).
9. ANDERSEN, A. Recent follow–up of nickel refinery workers in Norway and respiratory cancer. *In:* Nieboer, E. & Nriagu, J.O., ed. *Nickel and human health*. New York, Wiley, 1992, pp. 621–628.
10. ANDERSEN, A. ET AL. Exposure to nickel compounds and smoking in relation to incidence of lung and nasal cancer among nickel refinery workers. *Occupational and environmental medicine,* **53**: 708–713 (1996).

6.11 Platinum

Exposure evaluation

There is currently very little information on levels of exposure to soluble platinum compounds in the general environment, and there are no authenticated observations on adverse health effects in the population resulting from such exposure. The available data derived from air sampling and from dust deposition of total platinum are limited. Ambient air concentrations of platinum compounds that would occur in different scenarios have been estimated using dispersion models developed by the US Environmental Protection Agency *(1)*. Ambient air concentrations of total platinum in various urban exposure situations, assuming an average emission rate of approximately 20 ng/km from the monolithic three-way catalyst, were estimated. These concentrations are lower by a factor of 100 than those estimated for the old, pellet-type catalyst. In the exposure conditions studied, estimated ambient air concentrations of platinum ranged from 0.05 pg/m^3 to 0.09 ng/m^3. The WHO Task Group on Environmental Health Criteria for Platinum considered that environmental contamination with platinum from the monolithic three-way catalyst is likely to be very low or negligible *(2)*. The Group concluded that platinum-containing exhaust emissions from such catalysts most probably do not pose a risk for adverse health effects in the general population but it was recommended that, to be on the safe side, the possibility should be kept under review.

A recently completed pilot study sought to acquire information on direct and indirect sources and emissions of platinum group metals in the United Kingdom environment *(3)*. With regard to emissions from motor vehicle catalytic converters, samples of road dusts and soils were collected from areas with high and low traffic flows, for platinum and lead estimation. Higher levels of platinum were found in dusts and soils at major road intersections and on roads with high traffic densities, indicating traffic as the source of platinum at these sites *(4)*.

As platinum in road dust is at least partially soluble, it may enter the food chain so that diet may also be a major source of platinum intake in the non-industrially exposed population. This is suggested by the total diet study carried out in Australia in Sydney, an area of high traffic density, and Lord Howe Island, an area with very low traffic density. Blood platinum levels were similar in the two locations *(5)*.

In early studies on platinum-exposed workers, exposure levels were high. Values ranged from 0.9 to 1700 μg/m^3 in four British platinum refineries, giving rise to symptoms in 57% of the exposed workers *(6)*. Following the adoption of an occupational exposure limit with a threshold limit value (TLV) for soluble platinum salts of 2 μg/m^3 as an 8-hour time-weighted average, the incidence of platinum salt hypersensitivity has fallen, but sensitization in workers has still been observed. Thus, in a cross-sectional survey, skin sensitization was reported in 19% of 65 workers in a platinum refinery, where analysis of airborne dust showed levels of soluble platinum of 0.08–0.1 μg/m^3 in one department and less than 0.05 μg/m^3 in other areas *(7)*. In another plant with air levels generally below 0.08 μg/m^3, 20% of exposed workers were sensitized *(8)*. It is possible, however, that short, sharp exposures to concentrations above the TLV could have been responsible for some of these effects. In a 4-month study in a United States platinum refinery with a high prevalence of rhinitis and asthma, workplace concentrations exceeded the occupational limit of 2 μg/m^3 for 50–75% of the time *(9)*. The risk of developing platinum salt sensitivity appears to be correlated with exposure intensity, the highest incidence occurring in groups with the highest exposure, although no unequivocal concentration–effect relationship can be deduced from the reported studies.

Health risk evaluation

There is no convincing evidence for sensitization or for other adverse health effects following exposure to metallic platinum. Exposure to the halogenated platinum complexes already described has given rise to sensitization following occupational exposure to platinum concentrations in air greater than the TLV of 2 μg/m^3, and may have caused sensitization reactions at concentrations down to and even below the limit of detection in workplace monitoring of 0.05 μg/m^3. Furthermore, as subsequent exposure to minute concentrations of these platinum salts may lead to a recurrence of the health effects shown in Table 20 in previously sensitized subjects, it is not possible to define a no-effect level for these platinum compounds.

Because the correlation between platinum exposure concentration and the development of sensitization is unknown, the WHO Task Group *(2)* considered that a recommendation for a reduction in the occupational exposure limit cannot at present be justified. It did, however, recommend that the occupational exposure limit of 2 μg/m^3 be changed from an 8-hour time-weighted average to a ceiling value, and that personal sampling devices be used in conjunction with area sampling to determine more correctly the true platinum exposure. Should it be ascertained unequivocally that sensitization has occurred in workers consistently exposed to platinum

Table 20. Concentration–effect data for platinum

Concentration range	Average duration of exposure	Frequency of health effects in the general population	Health effects in susceptible groups
Airborne dust level for soluble platinum salts above the TLV time-weighted average of 2 μg/m³	Varies from weeks to years	No data available	In some occupationally exposed individuals: conjunctivitis, rhinitis, cough, wheeze, dyspnoea, asthma, contact dermatitis, urticaria, mucous membrane inflammation
Airborne dust level for soluble platinum salts < 0.05 μg/m³			Possibility that the above effects cannot be excluded Recurrence of the above effects in subjects previously sensitized Conversion to positive skin-prick test

levels below the current exposure limit of 2 μg/m³, and that intermittent, short exposures above this level had not taken place, there would be strong grounds for reducing the exposure limit.

The degree of solubilization and perhaps conversion to halide complexes of platinum particulate matter emitted into the general environment is not known, but is likely to be small. The prevalence of asthma in industrialized communities is increasing markedly. While there are no observations to suggest that platinum (emitted from vehicle catalytic converters or from industrial sources, deposited and in part converted in the general environment into halide salts) may act as an etiological agent, it would be inappropriate in the present state of knowledge to propose a no-effect level. From observations following occupational exposure, a value of 0.05 μg/m³ for soluble platinum salts may be considered as a tentative LOAEL. Platinum levels in air in the general environment are at least three orders of magnitude below this figure.

While *cis*-platin, an IARC Group 2A carcinogen, is released into the environment following medical use, there are no grounds for considering this platinum compound or its analogues as significant atmospheric pollutants.

Guidelines

In occupational settings, sensitization reactions have been observed for soluble platinum down to the limit of detection of 0.05 μg/m^3. However, these effects have occurred only in individuals previously sensitized by higher exposure levels. It is unlikely that the general population exposed to ambient concentrations of soluble platinum, which are at least three orders of magnitude lower, will develop similar effects. At present no specific guideline value is recommended but further studies are required, in particular on the speciation of platinum in the environment.

References

1. ROSNER, G. & MERGET, R. Allergenic potential of platinum compounds. *In:* Dayan, A.D. et al., ed. *Immunotoxicity of metals and immunotoxicology.* New York, Plenum Press, 1990, pp. 93–101.
2. *Platinum.* Geneva, World Health Organization, 1991 (Environmental Health Criteria, No. 125).
3. FARAGO, M..E. *Platinum group metals in the environment: their use in vehicle exhaust catalysts and implications for human health in the UK. A report prepared for the Department of the Environment.* London, IC Consultants, 1995.
4. FARAGO, M.E. ET AL. Platinum metal concentrations in urban road dust and soil in the United Kingdom. *Fresenius journal of analytical chemistry,* **354**: 660–663 (1996).
5. VAUGHAN, G.T. & FLORENCE, T.M. Platinum in the human diet, blood, hair and excreta. *Science of the total environment,* **111**: 47–58 (1992).
6. HUNTER, D. ET AL. Asthma caused by the complex salts of platinum. *British journal of industrial medicine,* **2**: 92–98 (1945).
7. BOLM-AUDORFF, U. ET AL. On the frequency of respiratory allergies in a platinum processing factory. *In:* Baumgartner, E. et al., ed. *Industrial change – occupational medicine facing new questions. Report of the 28th Annual Meeting of the German Society for Occupational Medicine, Innsbruck, Austria, 4–7 May, 1988.* Stuttgart, Gentner Verlag, 1988, pp. 411–416.
8. MERGET, R. ET AL. Asthma due to the complex salts of platinum – a cross sectional survey of workers in a platinum refinery. *Clinical allergy,* **18**: 569–580 (1988).
9. CALVERLEY, A.E. ET AL. Platinum salt sensitivity in refinery workers. Incidence and effects of smoking and exposure. *Occupational and environmental medicine,* **52**: 661–666 (1995).

6.12 Vanadium

Exposure evaluation

The natural background level of vanadium in air in Canada has been reported to be in the range 0.02–1.9 ng/m^3 *(1)*. Vanadium concentrations recorded in rural areas varied from a few nanograms to tenths of a nanogram per m^3, and in urban areas from 50 ng/m^3 to 200 ng/m^3. In cities during the winter, when fuel oil with a high vanadium content was used for heating, concentrations as high as 2000 ng/m^3 were reported. Air pollution by industrial plants may be less than that caused by power stations and heating equipment.

The concentrations of vanadium in workplace air (0.01–60 mg/m^3) are much higher than those in the general environment.

Health risk evaluation

The acute and chronic effects of vanadium exposure on the respiratory system of occupationally exposed workers should be regarded as the most significant factors when establishing air quality guidelines. Most of the clinical symptoms reported reflect irritative effects of vanadium on the upper respiratory tract, except at higher concentrations (above 1 mg vanadium per m^3), when more serious effects on the lower respiratory tract are observed. Clinical symptoms of acute exposure are reported *(2)* in workers exposed to concentrations ranging from 80 µg to several mg vanadium per m^3, and in healthy volunteers *(3)* exposed to concentrations of 56–560 µg/m^3 (Table 21).

A study of occupationally exposed groups provides data reasonably consistent with those obtained from controlled acute human exposure experiments, suggesting that the LOAEL for acute exposure can be considered to be 60 µg/m^3.

Chronic exposure to vanadium compounds revealed a continuum in the respiratory effects, ranging from slight changes in the upper respiratory tract, with irritation, coughing and injection of pharynx, detectable at 20 µg/m^3, to more serious effects such as chronic bronchitis and pneumonitis, which occurred at levels above 1 mg/m^3. Occupational studies illustrate the concentration–effect relationship at low levels of exposure *(4–6)*, showing increased prevalence of irritative symptoms of the upper respiratory tract; this suggests that 20 µg/m^3 can be regarded as the LOAEL for

Table 21. Respiratory effects after acute and chronic exposures to low levels of vanadium

Type of exposure	Vanadium compound	Concentration in µg/m³		Symptoms	Reference
		Compound	Vanadium		
Acute					
Boiler cleaning	V_2O_5 V_2O_3	523	80	Changes in parameters of lung functions	*(2)*
Clinical study (experimental 8-hour exposure)	V_2O_5	1000	560	Respiratory irritation: persistent and frequent cough, expiratory wheezes	*(3)*
	V_2O_5	200	112	Persistent cough (7–10 days)	
	V_2O_5	100	56	Slight cough for 4 days	
Chronic					
Vanadium refinery	V_2O_5	536	300	Respiratory irritation: cough, sputum, nose and throat irritation, injected pharynx	*(4)*
Vanadium refinery	V_2O_5	18–71	10–40	Irritative changes of mucous membranes of upper respiratory tract	*(5)*
Vanadium processing	V_2O_5 V_2O_3	—	1.2–12.0	Respiratory irritation: injected pharynx	*(6)*

chronic exposure (Table 21). There are no conclusive data on the health effects of exposure to airborne vanadium at present concentrations in the general population, and a susceptible subpopulation is not known. Vanadium is a potent respiratory irritant, however, which would suggest that asthmatics should be considered a special group at risk.

There are no well documented animal data to support findings in human studies, although one study reported systemic and local respiratory effects in rats at levels of 3.4–15 μg/m^3 *(7)*.

Guidelines

Available data from occupational studies suggest that the LOAEL of vanadium can be assumed to be 20 μg/m^3, based on chronic upper respiratory tract symptoms. Since the adverse nature of the observed effects on the upper respiratory tract were minimal at this concentration, and a susceptible subpopulation has not been identified, a protection factor of 20 was selected. It is believed that below 1 μg/m^3 (averaging time 24 hours) environmental exposure to vanadium is not likely to have adverse effects on health.

The available evidence indicates that the current vanadium levels generally found in industrialized countries are not in the range associated with potentially harmful effects.

References

1. COMMITTEE ON BIOLOGIC EFFECTS OF ATMOSPHERIC POLLUTANTS. *Vanadium.* Washington, DC, National Academy of Sciences, 1974.
2. LEES, R.E.M. Changes in lung function after exposure to vanadium compounds in fuel oil ash. *British journal of industrial medicine*, **37**: 253–256 (1980).
3. ZENZ, C. & BERG, B.A. Human responses to controlled vanadium pentoxide exposure. *Archives of environmental health,* **14**: 709–712 (1967).
4. LEWIS, C.E. The biological effects of vanadium. II. The signs and symptoms of occupational vanadium exposure. *AMA archives of industrial health,* **19**: 497–503 (1959).
5. KIVILUOTO, M. ET AL. Effects of vanadium on the upper respiratory tract of workers in a vanadium factory. *Scandinavian journal of work, environment and health,* **5**: 50–58 (1979).
6. NISHIYAMA, K. ET AL. [A survey of people working with vanadium pentoxide]. *Shikoku igaku zasshi,* **31**: 389–393 (1977) [Japanese].
7. PAZNYCH, V.M. [Maximum permissible concentration of vanadium pentoxide in the atmosphere]. *Gigiena i sanitarija,* **31**: 6–12 (1966) [Russian].

CHAPTER 7

Classical pollutants

7.1 Nitrogen dioxide

Exposure evaluation

Levels of nitrogen dioxide vary widely because a continuous baseline level is frequently present, with peaks of higher levels superimposed. Natural background annual mean concentrations are in the range 0.4–9.4 $\mu g/m^3$. Outdoor urban levels have an annual mean range of 20–90 $\mu g/m^3$ and hourly maxima in the range 75–1015 $\mu g/m^3$. Levels indoors where there are unvented gas combustion appliances may average more than 200 $\mu g/m^3$ over a period of several days. A maximum 1-hour peak may reach 2000 $\mu g/m^3$. For briefer periods, even higher concentrations have been measured.

Critical concentration–response data

Monotonic concentration–response data are available only from a few animal studies. Thus, this section will focus on lowest-observed-effect levels and their interpretation.

Short-term exposure effects

Available data from animal toxicology experiments rarely indicate the effects of acute exposure to nitrogen dioxide concentrations of less than 1880 $\mu g/m^3$ (1 ppm). Normal healthy people exposed at rest or with light exercise for less than 2 hours to concentrations of more than 4700 $\mu g/m^3$ (2.5 ppm) experience pronounced decrements in pulmonary function; generally, such people are not affected at less than 1880 $\mu g/m^3$ (1 ppm). One study showed that the lung function of people with chronic obstructive pulmonary disease is slightly affected by a 3.75-hour exposure to 560 $\mu g/m^3$ (0.3 ppm) *(1)*. A wide range of findings in asthmatics has been reported; one study observed no effects from a 75-minute exposure to 7520 $\mu g/m^3$ (4 ppm) *(2)*, whereas others showed decreases in FEV_1 after 10 minutes of exercise during exposure to 560 $\mu g/m^3$ (0.3 ppm) *(3)*.

Asthmatics are likely to be the most sensitive subjects, although uncertainties exist in the health database. The lowest concentration causing effects on pulmonary function was reported from two laboratories that exposed mild asthmatics for 30–110 minutes to 560 $\mu g/m^3$ (0.3 ppm) during intermittent exercise. However, neither of these laboratories was able to replicate these responses with a larger group of asthmatic subjects. One of these studies indicated that nitrogen dioxide can increase airway reactivity to cold air in asthmatics. At lower concentrations, the pulmonary function of asthmatics was not changed significantly.

Nitrogen dioxide increases bronchial reactivity as measured by pharmacological bronchoconstrictor agents in normal and asthmatic subjects, even at levels that do not affect pulmonary function directly in the absence of a bronchoconstrictor. Asthmatics appear to be more susceptible. For example, some but not all studies show increased responsiveness to bronchoconstrictors at nitrogen dioxide levels as low as 376–560 μg/m^3 (0.2–0.3 ppm); in other studies, higher levels had no such effect. Because the actual mechanisms are not fully defined and nitrogen dioxide studies with allergen challenges showed no effects at the lowest concentration tested (190 μg/m^3; 0.1 ppm), full evaluation of the health consequences of the increased responsiveness to bronchoconstrictors is not yet possible.

Long-term exposure effects

Studies with animals have clearly shown that several weeks to months of exposure to nitrogen dioxide concentrations of less than 1880 μg/m^3 (1 ppm) cause a plethora of effects, primarily in the lung but also in other organs, such as the spleen, liver and blood. Both reversible and irreversible lung effects have been observed. Structural changes range from a change in cell types in the tracheobronchial and pulmonary regions (lowest reported level 640 μg/m^3) to emphysema-like effects (at concentrations much higher than ambient). Biochemical changes often reflect cellular alterations (lowest reported levels for several studies 380–750 μg/m^3 (0.2–0.4 ppm) but isolated cases at lower effective concentrations). Nitrogen dioxide levels as low as 940 μg/m^3 (0.5 ppm) also increase susceptibility to bacterial and viral infection of the lung *(4)*.

There are no epidemiological studies that can be confidently used quantitatively to estimate long-term nitrogen dioxide exposure durations or concentrations likely to be associated with the induction of unacceptable health risks in children or adults. Because homes with gas cooking appliances have peak nitrogen dioxide levels that are in the same range as levels causing effects in some animal and human clinical studies, epidemiological studies evaluating the effects of nitrogen dioxide exposures in such homes have been of much interest. In general, epidemiological studies on adults and on infants under 2 years showed no significant effect of the use of gas cooking appliances on respiratory illness; nor do the few available studies of infants and adults show any associations between pulmonary function changes and gas stove use. However, children aged 5–12 years are estimated to have a 20% increased risk for respiratory symptoms and disease for each increase in nitrogen dioxide concentration of 28.3 μg/m^3 (2-week average) where the weekly average concentrations are in the range 15–128 μg/m^3 or possibly higher. Nevertheless, the observed effects cannot clearly be attributed to either the repeated short-term high-level peak exposures or long-term exposures in the range of the stated weekly averages (or possibly both).

As hinted at by the indoor studies, the results of outdoor studies tend to point consistently toward increased respiratory symptoms, their duration, and/or lung function decrements being qualitatively associated in children with long-term ambient nitrogen dioxide exposures. Outdoor epidemiology studies, as with indoor studies, however, provide little evidence for the association of long-term ambient exposures with health effects in adults. None of the available studies yields confident estimates of long-term exposure–effect levels, but available results are most clearly suggestive of respiratory effects in children at annual average nitrogen dioxide concentrations of 50–75 $\mu g/m^3$ or higher.

Health risk evaluation

Small, statistically significant, reversible effects on lung function and airway responsiveness have been observed in mild asthmatics during a 30-minute exposure to nitrogen dioxide concentrations of 380–560 $\mu g/m^3$ (0.2–0.3 ppm). The sequelae of repetitive exposures of such individuals or the impact of single exposures on more severe asthmatics are not known. In most animal experiments, however, 1–6 months of exposure to 560–940 $\mu g/m^3$ are required to produce changes in lung structure, lung metabolism and lung defences against bacterial infection. Thus, it is prudent to avoid exposures in humans, because repetitive exposures in animals lead to adverse effects. Animal toxicology studies of lung host defence and morphology suggest that peak concentrations contribute more to the toxicity of nitrogen dioxide than does duration, although duration is still important. Nitrogen dioxide puts children at increased risk of respiratory illness. This is of concern because repeated lung infections in children can cause lung damage later in life.

Nitrogen dioxide presents a dilemma with respect to guidelines. It is clear that the public should be protected from excessive exposure, but the recommendation of a guideline is complicated owing to the difficulties posed by the uncertainties in exposure–response relationships for both acute (< 3-hour) and long-term exposure, and the uncertainties in establishing an appropriate margin of protection. Studies of asthmatics exposed to 380–560 $\mu g/m^3$ indicate a change of about 5% in pulmonary function and an increase in airway responsiveness to bronchoconstrictors. Asthmatics are more susceptible to the acute effects of nitrogen dioxide: they have a higher baseline airway responsiveness. Thus, a nitrogen-dioxide-induced increase in airway responsiveness is expected to have clinical implications for exaggerated responses to a variety of provocative agents, such as cold air, allergies or exercise. Concern about asthmatics is also enhanced, considering the increase in the number of asthmatics in many countries (many countries have 4–6% asthmatics). A number of epidemiological studies of relatively large populations exposed indoors to peak levels of nitrogen dioxide from gas-combustion appliances have not provided consistent evidence of adverse pulmonary function effects. In one study, elderly women who used gas stoves had a high prevalence of asthma. Nevertheless, the human clinical studies of function and airway reactivity do not show monotonic concentration responses, and the studies are not internally consistent. Animal studies do not provide substantial evidence of biochemical, morphological or physiological effects in the lung following a single acute exposure to concentrations in the range of the lowest-observed-effect level in humans. On the other hand, the mild asthmatics chosen for the controlled exposure studies do not represent all asthmatics, and there are likely to be some individuals with greater sensitivity to nitrogen dioxide. Furthermore, subchronic and chronic animal studies do show significant morphological, biochemical and immunological changes.

The epidemiological studies discussed show increased risk of respiratory illness in children at an increase in nitrogen dioxide level of about 30 $\mu g/m^3$; most studies measured 2-week averages on personal samplers. It is not known, however, whether the effect was related to this 2-week average, the actual pattern (baseline and peaks) over the 2 weeks, the peaks over the 2 weeks, or some other index for a longer time-frame prior to the study measurement. It is also not possible to clearly discern the relative contributions of indoor and outdoor levels of nitrogen dioxide.

Guidelines

Despite the large number of acute controlled exposure studies on humans, several of which used multiple concentrations, there is no evidence for a clearly defined concentration–response relationship for nitrogen dioxide exposure. For acute exposures, only very high concentrations (1990 $\mu g/m^3$; > 1000 ppb) affect healthy people. Asthmatics and patients with chronic obstructive pulmonary disease are clearly more susceptible to acute changes in lung function, airway responsiveness and respiratory symptoms. Given the small changes in lung function (< 5% drop in FEV_1 between air and nitrogen dioxide exposure) and changes in airway responsiveness reported in several studies, 375–565 $\mu g/m^3$ (0.20–0.30 ppm) is a clear lowest-observed-effect level. A 50% margin of safety is proposed because of the reported statistically significant increase in response to a bronchoconstrictor (increased airway responsiveness) with exposure to 190 $\mu g/m^3$ and a meta-analysis suggesting changes in airway responsiveness below 365 μg/m^3. (The significance of the response at 190 μg/m^3 (100 ppb) has been questioned on the basis of an inappropriate statistical analysis.)

On the basis of these human clinical data, a 1-hour guideline of 200 μg/m^3 is proposed. At double this recommended guideline (400 μg/m^3) there is evidence to suggest possible small effects in the pulmonary function of asthmatics. Should the asthmatic be exposed either simultaneously or sequentially to nitrogen dioxide and an aeroallergen, the risk of an exaggerated response to the allergen is increased. At 50% of the suggested guideline (100 μg/m^3, 50 ppb) there have been no studies of acute response in 1 hour.

Although there is no particular study or set of studies that clearly support selection of a specific numerical value for an annual average guideline, the database nevertheless indicates a need to protect the public from chronic nitrogen dioxide exposure. For example, indoor air studies with a strong nitrogen dioxide source, such as gas stoves, suggest that an increment of about 30 μg/m^3 (2-week average) is associated with a 20% increase in lower respiratory illness in children aged 5–12 years. However, the affected children had a pattern of indoor exposure that included peak exposures higher than those typically encountered outdoors. Thus the results cannot be readily extrapolated quantitatively to the outdoor situation. Outdoor epidemiological studies have found qualitative evidence of ambient exposures being associated with increased respiratory symptoms and lung function decreases in children (most clearly suggestive at annual average concentrations of 50–75 μg/m^3 or higher and consistent with findings from indoor studies), although they do not provide clear exposure–response information for nitrogen dioxide. In these epidemiological studies, nitrogen dioxide has appeared to be a good indicator of the pollutant mixture. Furthermore, animal toxicological studies show that prolonged exposures can cause decreases in lung host defences and changes in lung structure. On these grounds, it is proposed that a long-term guideline for nitrogen dioxide be established. Selecting a well supported value based on the studies reviewed has not been possible, but it has been noted that a prior review conducted for the Environmental Health Criteria document on nitrogen oxides recommended an annual value of 40 μg/m^3 *(5)*. In the absence of support for an alternative value, this figure is recognized as an air quality guideline.

References

1. MORROW, P.E. & UTELL, M.J. *Responses of susceptible subpopulations to nitrogen dioxide*. Cambridge, MA, Health Effects Institute, 1989 (Research Report, No. 23).

2. LINN, W.S. & HACKNEY, J.D. *Short-term human respiratory effects of nitrogen dioxide: determination of quantitative dose–response profiles, phase II. Exposure of asthmatic volunteers to 4 ppm NO_2*. Atlanta, GA, Coordinating Research Council, Inc., 1984 (Report No. CRC-CAPM-48-83-02).
3. ROGER, L.J. ET AL. Pulmonary function, airway responsiveness, and respiratory symptoms in asthmatics following exercise in NO_2. *Toxicology and industrial health*, **6**: 155–171 (1990).
4. EHRLICH, R. & HENRY M.C. Chronic toxicity of nitrogen dioxide. I. Effect on resistance to bacterial pneumonia. *Archives of environmental health*, **17**: 860–865 (1968).
5. *Nitrogen oxides*. Geneva, World Health Organization, 1997 (Environmental Health Criteria, No. 188).

7.2 Ozone and other photochemical oxidants

Exposure evaluation

Ozone and other photochemical oxidants are formed by the action of short-wavelength radiation from the sun on nitrogen dioxide. In the presence of volatile organic compounds, the equilibrium favours the formation of higher levels of ozone. Background levels of ozone, mainly of anthropogenic origin, are in the range 40–70 μg/m³ (0.02–0.035 ppm) but can be as high as 120–140 μg/m³ (0.06–0.07 ppm) for 1 hour. In Europe, maximum hourly ozone concentrations may exceed 300 μg/m³ (0.15 ppm) in rural areas and 350 μg/m³ (0.18 ppm) in urbanized regions. Submaximal levels (80–90% of maximum) can occur for 8–12 hours a day for many consecutive days.

Health risk evaluation

Ozone toxicity occurs in a continuum in which higher concentrations, longer exposure duration and greater activity levels during exposure cause greater effects. Short-term acute effects include respiratory symptoms, pulmonary function changes, increased airway responsiveness and airway inflammation. These health effects were statistically significant at a concentration of 160 μg/m³ (0.08 ppm) for 6.6-hour exposures in a group of healthy exercising adults, with the most sensitive subjects experiencing functional decrements of > 10% within 4–5 hours *(1)*. Controlled exposures of heavily exercising adults or children to an ozone concentration of 240 μg/m³ (0.12 ppm) for 2 hours have also been observed to produce decrements in pulmonary function *(2, 3)*. There is no question that substantial acute adverse effects occur with 1 hour of exercising exposure at concentrations of 500 μg/m³ or higher, particularly in susceptible individuals or subgroups.

Field studies in children, adolescents and young adults have indicated that pulmonary function decrements can occur as a result of short-term exposure to ozone concentrations of 120–240 μg/m³ and higher. Mobile laboratory studies using ambient air containing ozone have observed associations between changes in pulmonary function in children or asthmatics and ozone concentrations of 280–340 μg/m³ (0.14–0.17 ppm) with exposures lasting several hours. Respiratory symptoms, especially cough, have been associated with ozone concentrations as low as 300 μg/m³ (0.15 ppm).

Ozone exposure has also been reported to be associated with increased hospital admissions for respiratory causes and exacerbation of asthma. That these effects are observed both with exposures to ambient ozone (and co-pollutants) and with controlled exposures to ozone alone demonstrates that the functional and symptomatic responses can be attributed primarily to ozone.

A number of studies evaluating rats and monkeys exposed to ozone for a few hours or days have shown alterations in the respiratory tract in which the lowest-observed-effect levels were in the range 160–400 μg/m^3 (0.08–0.2 ppm). These included the potentiation of bacterial lung infections, inflammation, morphological alterations in the lung, increases in the function of certain lung enzymes active in oxidant defences, and increases in collagen content. Long-term exposure to ozone in the range 240–500 μg/m^3 (0.12–0.25 ppm) causes morphological changes in the epithelium and interstitium of the centriacinar region of the lung, including fibrotic changes.

Guidelines

The selection of guidelines for ambient ozone concentrations is complicated by the fact that detectable responses occur at or close to the upper limits of background concentrations. At ozone levels of 200 μg/m^3 and lower (for exposure periods of 1–8 hours) there are statistically significant decrements in lung function, airway inflammatory changes, exacerbations of respiratory symptoms and symptomatic and functional exacerbations of asthma in exercising susceptible people. Functional changes and symptoms as well as increased hospital admissions for respiratory causes are also observed in population studies. Thus it is not possible to base the guidelines on a NOAEL or a LOAEL with an uncertainty factor of more than a small percentage. Thus, selection of a guideline has to be based on the premise that some detectable functional responses are of little or no health concern, and that the number of responders to effects of concern are too few to represent a group warranting protection from exposures to ambient ozone.

In the case of respiratory function responses, a judgement could be made that ozone-related reductions in FEV_1, for example, of < 10% were of no clinical concern. In the case of visits to clinics or emergency departments or hospital admissions for respiratory diseases, it would be necessary to determine how many cases per million population would be needed to constitute a group warranting societal protection. In the case of asthmatic children needing extra medication in response to elevated ozone concentrations, it would be necessary to conclude that medication will be available to sufficiently ameliorate their distress and thereby prevent more serious consequences.

On such a basis, a guideline value for ambient air of 120 μg/m³ for a maximum period of 8 hours per day is established as a level at which acute effects on public health are likely to be small.

For those public health authorities that cannot accept such levels of health risk, an alternative is to select explicitly some other level of acceptable exposure and associated risk. Tables 22 and 23 summarize the ambient ozone concentrations that are associated with specific levels of response among specified population subgroups. Although chronic exposure to ozone can cause effects, quantitative information from humans is inadequate to estimate the degree of protection from chronic effects offered by this guideline. In any case, the ozone concentration at which any adverse health outcome is expected will vary with the duration of the exposure and the volume of air that is inhaled during the exposure.

Thus, the amount of time spent outdoors and the typical level of activity are factors that should be considered in risk evaluation. Table 22 summarizes the ozone levels at which two representative adverse health outcomes,

Table 22. Health outcomes associated with controlled ozone exposures

Health outcome	Ozone concentration (μg/m³) at which the health effect is expected	
	Averaging time 1 hour	Averaging time 8 hours
Change in FEV_1 (active, healthy, outdoors, most sensitive 10% of young adults and children):		
5%	250	120
10%	350	160
20%	500	240
Increase in inflammatory changes (neutrophil influx) (healthy young adults at > 40 litres/minute outdoors)		
2-fold	400	180
4-fold	600	250
8-fold	800	320

Table 23. Health outcomes associated with changes in ambient ozone concentration in epidemiological studies

Health outcome	Change in ozone concentration (μg/m³)	
	Averaging time 1 hour	Averaging time 8 hours
Increase in symptom exacerbations among adults or asthmatics (normal activity):		
25%	200	100
50%	400	200
100%	800	300
Increase in hospital admissions for respiratory conditions:[a]		
5%	30	25
10%	60	50
20%	120	100

[a] Given the high degree of correlation between the 1-hour and 8-hour ozone concentration in field studies, the reduction in health risk associated with decreasing 1-hour or 8-hour ozone levels should be almost identical.

based on controlled exposure experiments, may be expected. The concentrations presented in this table have been established by experts on the basis of collective evidence from numerous studies and linear extrapolation in a few cases where data were limited.

Epidemiological data show relationships between changes in various health outcomes and changes in the peak daily ambient ozone concentration. Two examples of such relationships are shown in Table 23. Short-term increases in levels of ambient ozone are associated both with increased hospital admissions with a respiratory diagnosis and respiratory symptom exacerbations, both in healthy people and in asthmatics. These observations may be used to quantify expected improvements in health outcomes that may be associated with lowering the ambient ozone concentration. The values presented in the table assume a linear relationship between ozone concentration and health outcome. Uncertainties exist, however, concerning the forms of these relationships and it is unclear whether similar response slopes can be expected at widely different ambient ozone levels. In the event that such relationships are curvilinear (concave), the benefits of lowering the ozone concentration are likely to be greater when the average ambient level is higher. Consequently, if the ambient ozone concentration is already low, the benefits of lowering the concentration may be less than would be suggested by Table 23. Another important area of uncertainty is the degree to which other pollutants influence these relationships.

The first edition of *Air quality guidelines for Europe (4)* recommended a 1-hour guideline value of 150–200 μg/m³. Although recent research does not indicate that this guideline would necessarily be erroneous, the 8-hour guideline would protect against acute 1-hour exposures in this range and thus it is concluded that a 1-hour guideline is not necessary. Furthermore, the health problems of greatest concern (increased hospital admissions, exacerbations of asthma, inflammatory changes in the lung, and structural alterations in the lung) are more appropriately addressed by a guideline value that limits average daily exposure, and consequently inhaled dose and dose rate, rather than one designed to cover the rare short-duration deteriorations in air quality that may be associated with unusual meteorological conditions.

A guideline for peroxyacetyl nitrate is not warranted at present since it does not seem to pose a significant health problem at levels observed in the environment.

References

1. HORSTMAN, D.H. ET AL. Ozone concentration and pulmonary response relationships for 6.6-hour exposures with five hours of moderate exercise to 0.08, 0.10, and 0.12 ppm. *American review of respiratory disease*, **142**: 1158–1163 (1990).
2. MCDONNELL, W.F. ET AL. Pulmonary effects of ozone exposure during exercise: dose–response characteristics. *Journal of applied physiology: respiratory and environmental exercise physiology*, **54**: 1345–1352 (1983).
3. GONG, H. JR ET AL. Impaired exercise performance and pulmonary function in elite cyclists during low-level ozone exposure in a hot environment. *American review of respiratory disease*, **134**: 726–733 (1986).
4. *Air quality guidelines for Europe*. Copenhagen, WHO Regional Office for Europe, 1987 (WHO Regional Publications, European Series, No. 23).

7.3 Particulate matter

Exposure evaluation

Data on exposure levels to airborne inhalable particles are still limited for Europe. Data have mostly been obtained from studies not directly aimed at providing long-term distributions of exposure data for large segments of the population. Nevertheless, it seems that in northern Europe, PM_{10} levels (particulate matter in which 50% of particles have an aerodynamic diameter of less than 10 μm) are low, with winter averages even in urban areas not exceeding 20–30 μg/m^3. In western Europe, levels seem to be higher at 40–50 μg/m^3, with only small differences between urban and non-urban areas. Levels in some central and eastern European locations from which data are available appear nowadays to be only a little higher than those measured in cities such as Amsterdam and Berlin. As a result of the normal day-to-day variation in PM_{10} concentrations, 24-hour averages of 100 μg/m^3 are regularly exceeded in many areas in Europe, especially during winter inversions.

Health risk evaluation

A variety of methods exist to measure particulate matter in air. For the present evaluation, studies have been highlighted in which particulate matter exposure was expressed as the thoracic fraction (~ PM_{10}) or size fractions or constituents thereof. Practically speaking, at least some data are also available on fine particles ($PM_{2.5}$), sulfates and strong aerosol acidity. Health effect studies conducted with (various forms of) total suspended particulates or black smoke as exposure indicators have provided valuable additional information in recent years. They are, however, less suitable for the derivation of exposure–response relationships for particulate matter, because total suspended particulates include particles that are too large to be inhaled or because the health significance of particle opacity as measured by the black smoke method is uncertain.

Recent studies suggest that short-term variations in particulate matter exposure are associated with health effects even at low levels of exposure (below 100 μg/m^3). The current database does not allow the derivation of a threshold below which no effects occur. This does not imply that no threshold exists; epidemiological studies are unable to define such a threshold, if it exists, precisely.

At low levels of (short-term) exposure (defined as 0–100 μg/m^3 for PM_{10}), the exposure–response curve fits a straight line reasonably well. There are indications from studies conducted in the former German Democratic Republic and in China, however, that at higher levels of exposure (several hundreds of $\mu g/m^3$ PM_{10}) the curve is shallower, at least for effects on mortality. In the London mortality studies, there was also evidence of a curvilinear relationship between black smoke and daily mortality, the slope becoming shallower at higher levels of exposure. Estimates of the magnitude of effect occurring at low levels of exposure should therefore not be used to extrapolate to higher levels outside the range of exposures that existed in most of the recent acute health effect studies.

Although there are now many studies showing acute effect estimates of PM_{10} that are quantitatively reasonably consistent, this does not imply that particle composition or size distribution within the PM_{10} fraction is unimportant. Limited evidence from studies on dust storms indicates that such PM_{10} particles are much less toxic than those associated with combustion sources. Recent studies in which PM_{10} size fractions and/or constituents have been measured suggest that the observed effects of PM_{10} are in fact largely associated with fine particles, strong aerosol acidity or sulfates (which may serve as a proxy for the other two) and not with the coarse (PM_{10} minus $PM_{2.5}$) fraction.

Traditionally, particulate matter air pollution has been thought of as a primarily urban phenomenon. It is now clear that in many areas of Europe, urban–rural differences in PM_{10} are small or even absent, indicating that particulate matter exposure is widespread. Indeed, several of the health effect studies reviewed in this chapter were conducted in rural or semirural rather than urban areas. This is not to imply that exposure to primary, combustion-related particulate matter may not be higher in urban areas. At present, however, data are lacking on the specific health risks of such exposures.

Evidence is emerging also that long-term exposure to low concentrations of particulate matter in air is associated with mortality and other chronic effects, such as increased rates of bronchitis and reduced lung function. Two cohort studies conducted in the United States suggest that life expectancy may be shortened by more than a year in communities exposed to high concentrations compared to those exposed to low concentrations. This is consistent with earlier results from cross-sectional studies comparing age-adjusted mortality rates across a range of long-term average concentrations. Again, such effects have been suggested to be associated with long-term average exposures that are low, starting at a concentration of fine particulate matter of about 10 $\mu g/m^3$. Whereas such observations require further corroboration, preferably also from other areas in the world, these new studies suggest that the public health implications of particulate matter exposure may be large.

Evaluation of the effects of short-term exposure on mortality and morbidity

Table 24 shows the summary estimates of relative increase in daily mortality, respiratory hospital admissions, reporting of bronchodilator use, cough and lower respiratory symptoms, and changes in peak expiratory flow associated with a 10 μg/m^3 increase in PM_{10} or $PM_{2.5}$, as reported in studies in which PM_{10} and/or $PM_{2.5}$ concentrations were actually measured (as opposed to being inferred from other measures such as coefficient of haze, black smoke or total suspended particulates). The database for parameters other than PM_{10} is still limited, but for the reasons noted above, it is very important to state that even though the evaluation of (especially the short-term) health effects is largely expressed in terms of PM_{10}, future regulations and monitoring activities should place emphasis on (appropriate representations of) the respiratory fraction in addition to, or even preferred to, PM_{10} *(1)*.

It is important to realize that at present it is not known what reduction in life expectancy is associated with daily mortality increases related to particulate matter exposure. If effects are restricted to people in poor health, effects on age at death may be small.

Table 24. Summary of relative risk estimates for various endpoints associated with a 10 μg/m^3 increase in the concentration of PM_{10} or $PM_{2.5}$

Endpoint	Relative risk for $PM_{2.5}$ (95% confidence interval)	Relative risk for PM_{10} (95% confidence interval)
Bronchodilator use		1.0305 (1.0201–1.0410)
Cough		1.0356 (1.0197–1.0518)
Lower respiratory symptoms		1.0324 (1.0185–1.0464)
Change in peak expiratory flow (relative to mean)		–0.13% (–0.17% to –0.09%)
Respiratory hospital admissions		1.0080 (1.0048–1.0112)
Mortality	1.015 (1.011–1.019)	1.0074 (1.0062–1.0086)

The effect estimates in Table 24 can be used with considerable reservation to estimate, for a population of a given size and mortality and morbidity experience, how many people would be affected over a short period of time with increased particulate matter levels. The reservation stems from the finding that for some of the estimated effects, there was no evidence of heterogeneity between studies in the magnitude of the effect estimate. An investigation of the reasons for heterogeneity is beyond the scope of this chapter. As a consequence, the pooled effect estimates may not be applicable in all possible circumstances.

For illustrative purposes, Table 25 contains an estimate of the effect of a 3-day episode with daily PM_{10} concentrations averaging 50 µg/m^3 and 100 µg/m^3 on a population of 1 million people. Table 25 makes it clear that, in a population of that size, the number of people dying or having to be admitted to hospital as a result of particulate matter exposure is small relative to the additional number of "person-days" of increased medication use and/or increased respiratory symptoms due to exposure to particulate matter.

Whereas these calculations should be modified according to the size, mortality and morbidity experience of populations of interest and, where possible, for factors contributing to the heterogeneity in the effect estimates, they do provide some insight into the public health consequences of certain exposures to particulate matter.

Table 25. Estimated number of people (in a population of 1 million) experiencing health effects over a period of 3 days characterized by a mean PM_{10} concentration of 50 or 100 µg/m^3

Health effect indicator	No. of people affected by a three-day episode of PM_{10} at:	
	50 µg/m^3	100 µg/m^3
No. of deaths	4	8
No. of hospital admissions due to respiratory problems	3	6
Person-days of bronchodilator use	4 863	10 514
Person-days of symptom exacerbation	5 185	11 267

Evaluation of the effects of long-term exposure on mortality and morbidity

The most convincing information on long-term effects of particulate matter exposure on mortality is provided by two recent cohort studies. Relative risk estimates for total mortality from the first study *(2)*, expressed per 10 μg/m^3, were 1.10 for inhalable particles (measured as either PM_{15} or PM_{10}), 1.14 for fine particles ($PM_{2.5}$) and 1.33 for sulfates. Relative risk estimates for total mortality from the second study *(3)*, expressed per 10 μg/m^3, were 1.07 for fine particles ($PM_{2.5}$) and 1.08 for sulfates. Sulfate levels used in the second study (range 3.6–23.6 μg/m^3) may have been inflated owing to sulfate formation on filter material used in earlier studies. The first study included one of the high-sulfate communities (Steubenville), yet the range of sulfate levels in this study was much lower (4.8–12.8 μg/m^3), possibly owing to the more adequate measurement methods employed in this study.

Long-term effects of particulate matter exposure on morbidity have been demonstrated in the Harvard 24 cities study among children *(4, 5)*. Expressed per 10 μg/m^3, the relative risks for bronchitis were 1.34 for $PM_{2.1}$, 1.29 for PM_{10}, and 1.96 for sulfate particles. The corresponding changes in FEV_1 were –1.9% ($PM_{2.1}$), –1.2% (PM_{10}) and –3.1% (sulfate particles). Whereas such mean changes are clinically unimportant, the proportion of children having a clinically relevant reduced lung function (forced vital capacity (FVC) or FEV_1 < 85% of predicted) was increased by a factor of 2–3 across the range of exposures *(5)*. A recent study from Switzerland *(6)* has shown significant reductions in FEV_1 of –1.0% per 10 μg/m^3 PM_{10}.

Table 26 provides a summary of the current knowledge of effects of long-term exposure to particulate matter on morbidity and mortality endpoints.

Using the risk estimates presented in Table 26, Table 27 provides estimates of the number of people experiencing health effects associated with long-term exposure to particulate matter, using similar assumptions about population size and morbidity as in Table 25. Specifically, a population size of one million has been assumed, 20% of whom are children, with a baseline prevalence of 5% for bronchitis symptoms among children (that is, 10 000 children are assumed to have bronchitis symptoms) and with a baseline prevalence of 3% of children (6000 children) having a lung function (FVC or FEV_1) lower than 85% of predicted.

In addition, the impact of long-term exposures to particulate matter on total mortality can be estimated. The number of persons surviving to a

Table 26. Summary of relative risk estimates for effects of long-term exposure to particulate matter on the morbidity and mortality associated with a 10 μg/m³ increase in the concentration of $PM_{2.5}$ or PM_{10}

Endpoint	Relative risk for $PM_{2.5}$ (95% confidence interval)	Relative risk for PM_{10} (95% confidence interval)
Death *(2)*	1.14 (1.04–1.24)	1.10 (1.03–1.18)
Death *(3)*	1.07 (1.04–1.11)	
Bronchitis *(4)*	1.34 (0.94–1.99)	1.29 (0.96–1.83)
Percentage change in FEV_1, children *(5)* [a]	−1.9% (−3.1% to −0.6%)	−1.2% (−2.3% to −0.1%)
Percentage change in FEV_1, adults *(6)*		−1.0% (not available)

[a] For $PM_{2.1}$ rather than $PM_{2.5}$.

Table 27. Estimated number of children (out of 200 000 in a population of 1 million) experiencing health effects per year due to long-term exposure to a $PM_{2.5}$ concentration of 10 or 20 μg/m³ above a background level of 10 μg/m³

Health effect indicator	No. of children affected per year at $PM_{2.5}$ concentrations above background of:	
	10 μg/m³	20 μg/m³
No. of additional children with bronchitis symptoms	3350	6700
No. of additional children with lung function (FVC or FEV_1) below 85% of predicted	4000	8000

certain age will be smaller in a population exposed to higher concentrations, and the difference will depend on the age group. If the mortality structure of Dutch males is taken as a basis for calculation, and if the assumptions used in the construction of Table 25 are applied, in each birth cohort of 100 000 men the number of survivors exposed to pollution increased by 10 μg/m³ (PM_{10}) will be reduced by 383 men before the age of 50, by 1250 men before the age of 60 and by 3148 men before the age of 70. An increase in the long-term exposure of 20 μg/m^3 (PM_{10}) corresponds to an estimated reduction of the number of men surviving to a certain age in the cohorts by, respectively, 764, 2494 or 6250 men.

Guidelines

The weight of evidence from numerous epidemiological studies on short-term responses points clearly and consistently to associations between concentrations of particulate matter and adverse effects on human health at low levels of exposure commonly encountered in developed countries. The database does not, however, enable the derivation of specific guideline values at present. Most of the information that is currently available comes from studies in which particles in air have been measured as PM_{10}. There is now also a sizeable body of information on fine particulate matter ($PM_{2.5}$) and the latest studies are showing that this is generally a better predictor of health effects than PM_{10}. Evidence is also emerging that constituents of $PM_{2.5}$ such as sulfates are sometimes even better predictors of health effects than $PM_{2.5}$ *per se*.

The large body of information on studies relating day-to-day variations in particulate matter to day-to-day variations in health provides quantitative estimates of the effects of particulate matter that are generally consistent. The available information does not allow a judgement to be made of concentrations below which no effects would be expected. Effects on mortality, respiratory and cardiovascular hospital admissions and other health variables have been observed at levels well below 100 μg/m^3, expressed as a daily average PM_{10} concentration. For this reason, no guideline value for short-term average concentrations is recommended either. Risk managers are referred to the risk estimates provided in the tables for guidance in decision-making regarding standards to be set for particulate matter.

The body of information on long-term effects is still smaller. Some studies have suggested that long-term exposure to particulate matter is associated with reduced survival, and a reduction of life expectancy in the order of 1–2 years. Other recent studies have shown that the prevalence of bronchitis symptoms in children, and of reduced lung function in children and adults, are associated with particulate matter exposure. These effects have been observed at annual average concentration levels below 20 μg/m^3 (as $PM_{2.5}$) or 30 μg/m^3 (as PM_{10}). For this reason, no guideline value for long-term average concentrations is recommended. Risk managers are referred to the risk estimates provided in the tables for guidance in decision-making regarding standards to be set for particulate matter.

References

1. LIPPMANN, M. & THURSTON, G.D. Sulphate concentrations as an indicator of ambient particulate matter air pollution for health risk evaluations. *Journal of exposure analysis and environmental epidemiology*, **6**: 123–146 (1996).
2. DOCKERY, D.W. ET AL. An association between air pollution and mortality in six U.S. cities. *New England journal of medicine*, **329**: 1753–1759 (1993).
3. POPE, C.A. III. ET AL. Particulate air pollution as a predictor of mortality in a prospective study of U.S. adults. *American journal of respiratory and critical care medicine*, **151**: 669–674 (1995).
4. DOCKERY, D.W. ET AL. Health effects of acid aerosols on North American children: respiratory symptoms. *Environmental health perspectives*, **104**: 500–505 (1996).
5. RAIZENNE, M. ET AL. Health effects of acid aerosols on North American children: pulmonary function. *Environmental health perspectives*, **104**: 506–514 (1996).
6. ACKERMANN-LIEBRICH, U. ET AL. Lung function and long-term exposure to air pollutants in Switzerland. *American journal of respiratory and critical care medicine*, **155**: 122–129 (1997).

7.4 Sulfur dioxide

Exposure evaluation

In much of western Europe and North America, concentrations of sulfur dioxide in urban areas have continued to decline in recent years as a result of controls on emissions and changes in fuel use. Annual mean concentrations in such areas are now mainly in the range 20–60 μg/m^3 (0.007–0.021 ppm), with daily means seldom more than 125 μg/m^3 (0.044 ppm). In large cities where coal is still widely used for domestic heating or cooking, however, or where there are poorly controlled industrial sources, concentrations may be 5–10 times these values. Peak concentrations over shorter averaging periods, of the order of 10 minutes, can reach 1000–2000 μg/m^3 (0.35–0.70 ppm) in some circumstances, such as the grounding of plumes from major point sources or during peak dispersion conditions in urban areas with multiple sources.

Health risk evaluation

Short-term exposures (less than 24 hours)

The most direct information on the acute effects of sulfur dioxide comes from controlled chamber experiments on volunteers. Most of these studies have been for exposure periods ranging from a few minutes up to 1 hour, but the exact duration is not critical because responses occur very rapidly, within the first few minutes after commencement of inhalation; continuing the exposure further does not increase effects *(1–3)*. The effects observed include reductions in FEV_1 or other indices of ventilatory capacity, increases in specific airway resistance, and symptoms such as wheezing or shortness of breath. Such effects are enhanced by exercise, which increases the volume of air inspired thereby allowing sulfur dioxide to penetrate further into the respiratory tract *(4, 5)*.

A wide range of sensitivity has been demonstrated, both among normal individuals and among those with asthma, who form the most sensitive group *(1, 4, 6, 7)*. Continuous exposure–response relationships, without any clearly defined threshold, are evident. To develop a guideline value, the minimum concentrations associated with adverse effects in the most extreme circumstances, that is with asthmatic patients exercising in chambers, have been considered. An example of an exposure–response relationship for such subjects, expressed in terms of reductions in FEV_1 after a 15-minute exposure, comes from a study by Linn et al. (*8*). Only small changes, not regarded as of clinical significance, were seen at 572 μg/m^3 (0.2 ppm); reductions representing about 10% of baseline FEV_1 occurred at about 1144 μg/m^3 (0.4 ppm); and reductions of about 15% occurred at about 1716 μg/m^3 (0.6 ppm). The response was not greatly influenced by the severity of asthma. These findings are consistent with those reported from other exposure studies. In one early series, however, a small change in airway resistance was reported in two of the asthmatic patients at 286 μg/m^3 (0.1 ppm).

Exposure over a 24-hour period

Information on effects of exposure averaged over a 24-hour period is derived mainly from epidemiological studies in which the effects of sulfur dioxide, particulate matter and other associated pollutants are considered *(9)*. Exacerbation of symptoms among panels of selected sensitive patients occurred consistently when the sulfur dioxide concentration exceeded 250 μg/m^3 (0.087ppm) in the presence of particulate matter. Such findings have related mainly to situations in which emissions from the inefficient burning of coal in domestic appliances have been the main contributor to the pollution complex. Several more recent studies, involving the mixed industrial and vehicular sources that now dominate, have consistently demonstrated effects on mortality (total, cardiovascular and respiratory) *(10–18)* and hospital emergency admissions *(14, 19–22)* for total respiratory causes and chronic obstructive pulmonary disease at lower levels of exposure (mean annual levels below 50 μg/m^3; daily levels usually not exceeding 125 μg/m^3). These results have been shown, in some instances, to persist when levels of black smoke and total suspended particulate matter were controlled for, while in other studies no attempts were made to separate the effects of the pollutants. No obvious threshold levels could so far be identified in those studies.

Long-term exposure

A similar situation arises in respect of effects of long-term exposures, expressed as annual averages. Earlier assessments examined findings on the prevalence of respiratory symptoms, respiratory illness frequencies, or differences in lung function values in localities with contrasting concentrations of sulfur dioxide and particulate matter, largely in the coal-burning era. The LOAEL of sulfur dioxide was judged to be 100 μg/m^3 (0.035 ppm) annual average, together with particulate matter. More recent studies related to industrial sources, or to the changed urban mixture, have shown adverse effects below this level, but a major difficulty in interpretation is that long-term effects are liable to be affected not only by current conditions but also by the qualitatively and quantitatively different pollution of earlier years. Cohort studies of differences in mortality between areas with contrasting pollution levels indicate that there is a closer association with particulate matter than with sulfur dioxide *(23, 24).*

Guidelines

Short-term exposures

Controlled studies with exercising asthmatics indicate that some asthmatics experience changes in pulmonary function and respiratory symptoms after periods of exposure as short as 10 minutes. Based on this evidence, it is recommended that a value of 500 μg/m^3 (0.175 ppm) should not be exceeded over averaging periods of 10 minutes. Because exposure to sharp peaks depends on the nature of local sources, no single factor can be applied to this value in order to estimate corresponding guideline values over somewhat longer periods, such as an hour.

Exposure over a 24-hour period and long-term exposure

Day-to-day changes in mortality, morbidity or lung function related to 24-hour average concentrations of sulfur dioxide are necessarily based on epidemiological studies in which people are in general exposed to a mixture of pollutants, which is why guideline values for sulfur dioxide have previously been linked with corresponding values for particulate matter. This approach led to a previous guideline value of 125 μg/m^3 (0.04 ppm) as a 24-hour average, after applying an uncertainty factor of 2 to the LOAEL. In more recent studies, adverse effects with significant public health importance have been observed at much lower levels of exposure. Nevertheless, there is still uncertainty as to whether sulfur dioxide is the pollutant responsible for the observed adverse effects or, rather, a surrogate for ultrafine particles or some other correlated substance. There is no basis for revising the 1987 guidelines for sulfur dioxide *(9)* and thus the following guidelines are recommended:

24 hours:	125 μg/m^3
annual:	50 μg/m^3

It should be noted that, unlike in the 1987 guidelines, these values for sulfur dioxide are no longer linked with particles.

References

1. Lawther, P.J. et al. Pulmonary function and sulphur dioxide: some preliminary findings. *Environmental research*, **10**: 355–367 (1975).

2. SHEPPARD, D. ET AL. Exercise increases sulfur dioxide induced bronchoconstriction in asthmatic subjects. *American review of respiratory disease*, **123**: 486–491 (1981).
3. LINN, W.S. ET AL. Asthmatics responses to 6-hr sulfur dioxide exposures on two successive days. *Archives of environmental health*, **39**: 313–319 (1984).
4. DEPARTMENT OF HEALTH. *Advisory Group on the Medical Aspects of Air Pollution Episodes. Second report: sulphur dioxide, acid aerosols and particulates*. London, H.M. Stationery Office, 1992.
5. BETHEL R.A. ET AL. Effect of exercise rate and route of inhalation on sulfur dioxide induced bronchoconstriction in asthmatic subjects. *American review of respiratory disease*, **128**: 592–596 (1983).
6. NADEL, J.A. ET AL. Mechanism of bronchoconstriction during inhalation of sulfur dioxide. *Journal of applied physiology*, **20**: 164–167 (1965).
7. HORSTMAN, D.H. ET AL. The relationship between exposure duration and sulphur dioxide induced bronchoconstriction in asthmatic subjects. *American Industrial Hygiene Association journal*, **49**: 38–47 (1988).
8. LINN, W.S. ET AL. Replicated dose–response study of sulfur dioxide effects in normal, atopic and asthmatic volunteers. *American review of respiratory disease*, **136**: 1127–1134 (1987).
9. *Air quality guidelines for Europe*. Copenhagen, WHO Regional Office for Europe, 1987 (WHO Regional Publications, European Series, No. 23).
10. SPIX, C. ET AL. Air pollution and daily mortality in Erfurt, East Germany, 1980–1989. *Environmental health perspectives*, **101**: 518–526 (1993).
11. WIETLISBACH, V. ET AL. Air pollution and daily mortality in three Swiss urban areas. *Social and preventive medicine*, **41**: 107–115 (1996).
12. SCHWARTZ, J. & DOCKERY, D.W. Increased mortality in Philadelphia associated with daily air pollution concentrations. *American review of respiratory disease*, **145**: 600–604 (1992).
13. SUNYER, J. ET AL. Air pollution and mortality in Barcelona. *Journal of epidemiology and community health*, **50** (Suppl.): S76–S80 (1996).
14. DAB, W. ET AL. Short-term respiratory health effects of ambient air pollution: results of the APHEA project in Paris. *Journal of epidemiology and community health*, **50** (Suppl.): S42–S46 (1996).
15. ZMIROU, D. ET AL. Short-term effects of air pollution on mortality in the city of Lyons, France 1985–1990. *Journal of epidemiology and community health*, **50** (Suppl.): S30–S35 (1996).
16. TOULOUMI, G. ET AL. Daily mortality and air pollution from particulate matter, sulphur dioxide and carbon monoxide in Athens, Greece: 1987–1991. A time-series analysis within the APHEA project. *Journal of epidemiology and community health*, **50** (Suppl.): S47–S51 (1996).

17. ANDERSON, H.R. ET AL. Air pollution and daily mortality in London: 1987–92. *British medical journal*, **312**: 665–669 (1996).
18. KATSOUYANNI, K. ET AL. Short-term effects of air pollution on health: a European approach using epidemiologic time series data. *European respiratory journal*, **8**: 1030–1038 (1995).
19. SCHWARTZ, J. & MORRIS, R. Air pollution and hospital admissions for cardiovascular disease in Detroit, Michigan. *American journal of epidemiology*, **142**: 23–35 (1995).
20. SUNYER, J. ET AL. Air pollution and emergency room admissions for chronic obstructive pulmonary diseases. *American journal of epidemiology*, **134**: 277–286 (1991).
21. PONCE DE LEON, A. ET AL. The effects of air pollution on daily hospital admission for respiratory disease in London:1987–88 to 1991–92. *Journal of epidemiology and community health*, **50** (Suppl.): S63–S70 (1996).
22. SCHOUTEN, J.P. ET AL. Short-term effects of air pollution on emergency hospital admissions for respiratory disease: results of the APHEA project in two major cities in the Netherlands, 1977–89. *Journal of epidemiology and community health*, **50** (Suppl.): S22–S29 (1996).
23. DOCKERY, D.W. ET AL. An association between air pollution and mortality in six U.S. cities. *New England journal of medicine*, **329**: 1753–1759 (1993).
24. POPE, C.A. III. ET AL. Particulate air pollution as a predictor of mortality in a prospective study of U.S. adults. *American journal of respiratory and critical care medicine*, **151**: 669–674 (1995).

CHAPTER 8

Indoor air pollutants

8.1 Environmental tobacco smoke

Exposure evaluation

Environmental tobacco smoke (ETS) is a dynamic complex mixture of thousands of compounds in particulate and vapour phases, and cannot be measured directly as a whole. Instead, various marker compounds, such as nicotine and respirable suspended particulates (RSPs), are used to quantify environmental exposure. In the United States, nicotine concentrations in homes where smoking occurs typically range from less than 1 μg/m^3 to over 10 μg/m^3 (*1*). Concentrations in offices where people smoke typically range from near zero to over 30 μg/m^3. Levels in restaurants, and especially bars, tend to be even higher, and concentrations in confined spaces such as cars can be higher still. Measurements of ETS-associated RSPs in homes where people smoke range from a few μg/m^3 to over 500 μg/m^3, while levels in offices are generally less than 100 μg/m^3 and those in restaurants can exceed 1 mg/m^3. ETS levels are directly related to smoker density; in countries with a higher smoking prevalence, average ETS levels could be higher.

In Western societies, with adult smoking prevalences of 30–50%, it is estimated that over 50% of homes are occupied by at least one smoker, resulting in a high prevalence of ETS exposure in children and other nonsmokers. A large percentage of nonsmokers are similarly exposed at work.

Health risk evaluation

ETS has been shown to increase the risks for a variety of health effects in nonsmokers exposed at typical environmental levels. The pattern of health effects from ETS exposure produced in adult nonsmokers is consistent with the effects known to be associated with active cigarette smoking. Chronic exposures to ETS increase lung cancer mortality *(1–5)*. In addition, the combined evidence from epidemiology and studies of mechanisms leads to the conclusion that ETS increases the risk of morbidity and mortality from cardiovascular disease in nonsmokers, especially those with chronic exposure *(4, 6–11)*. ETS also irritates the eyes and respiratory tract. In infants and young children, ETS increases the risk of pneumonia, bronchitis, bronchiolitis and fluid in the middle ear *(1, 2, 13, 14)*. In asthmatic children, ETS increases the severity and frequency of asthma attacks *(12)*. Furthermore, as with active smoking, ETS reduces birth weight in the offspring of nonsmoking mothers *(15)*.

Other health effects have also been associated with ETS exposure, but the evidence is not as conclusive. In adults, there is strong suggestive evidence that ETS increases mortality from sinonasal cancer *(16, 17)*. In infants, recent evidence suggests that ETS is a risk factor for sudden infant death syndrome *(18–22)*.

Populations at special risk for the adverse health effects of ETS are young children and infants, asthmatics, and adults with other risk factors for cardiovascular disease. Levels of exposure where these effects have been observed are indicated by nicotine levels of 1–10 $\mu g/m^3$ (nicotine has been demonstrated to be a reliable marker of ETS levels).

Because of the extensive prevalence of ETS exposure and the high incidence of some of the health effects associated with ETS exposure, such as cardiovascular disease in adults and lower respiratory tract infections in children, even small increases in relative risks can translate into substantial levels of mortality and morbidity on a population basis.

Based on the combined evidence from several studies, WHO has estimated that some 9–13% of all cancer cases can be attributed to ETS in a nonsmoking population of which 50% are exposed to ETS. The proportion of lower respiratory illness in infants attributed to ETS exposure can be estimated at 15–26%, assuming that 35% of the mothers smoke at home. Those estimates, when applied to the European population, will result in approximately 3000–4500 cases of cancer in adults per year, and between 300 000 and 550 000 episodes of lower respiratory illness in infants per year, which are expected to be related to ETS exposure *(23)*.

Comparable results were calculated for nonsmokers in the United States *(1)*. The US Environmental Protection Agency (EPA) recently estimated that ETS causes 3000 lung cancer deaths in adult nonsmokers (roughly 100 million people who have never smoked and long-term former smokers) in the United States each year. The EPA also estimated that ETS is responsible for between 150 000 and 300 000 lower respiratory tract infections annually in the roughly 5.5 million children under 18 months of age, and that it exacerbates asthma in about 20% of asthmatic children. These estimates are based on a large quantity of human data from actual exposure levels, and involve no high-to-low-dose or animal-to-human extrapolations; thus confidence in these estimates is considered high.

Quantitative population estimates for cardiovascular disease mortality are less certain than those for lung cancer. The main reasons for greater quantitative uncertainty in estimates for cardiovascular disease are that (*a*) there are fewer epidemiological data available (in particular, there are few data for males, which is especially critical because males have a very different baseline risk of cardiovascular disease than females), and (*b*) there are more risk factors for cardiovascular disease that need to be adjusted for to obtain a reliable risk estimate. In general, the relative risk estimates for cardiovascular disease from ETS exposure are similar to those for lung cancer; however, the baseline risk of death from cardiovascular disease in nonsmokers is at least 10 times higher than the risk of lung cancer. Therefore, the population risks could be roughly 10 times higher as well. Thus, while there is more confidence in the presented estimates for lung cancer, the public health impact of ETS is expected to be substantially greater for cardiovascular disease.

Guidelines

ETS has been found to be carcinogenic in humans and to produce a substantial amount of morbidity and mortality from other serious health effects at levels of 1–10 μg/m^3 nicotine (taken as an indicator of ETS). Acute and chronic respiratory health effects on children have been demonstrated in homes with smokers (nicotine 1–10 μg/m^3) and even in homes with occasional smoking (0.1–1 μg/m^3). There is no evidence for a safe exposure level. The unit risk of cancer associated with lifetime ETS exposure in a home where one person smokes is approximately 1×10^{-3}.

References

1. *Respiratory health effects of passive smoking: lung cancer and other disorders*. Washington, DC, US Environmental Protection Agency, 1992 (EPA/600/6-90/006F).
2. *The health consequences of involuntary smoking. A report of the Surgeon General.* Washington, DC, US Department of Health and Human Services, 1986 (DHHS Publication No. (PHS) 87–8398).
3. *Tobacco smoking*. Lyons, International Agency for Research on Cancer, 1986 (IARC Monographs on the Evaluation of the Carcinogenic Risk of Chemicals to Humans, Vol. 38).
4. *Environmental tobacco smoke in the workplace: lung cancer and other health effects*. Cincinnati, OH, National Institute for Occupational Safety and Health, 1991 (Current Intelligence Bulletin, No. 54).
5. FONTHAM, E.T.H. ET AL. Environmental tobacco smoke and lung cancer in nonsmoking women: a multicenter study. *Journal of the American Medical Association*, **271**: 1752–1759 (1994).
6. GLANTZ, S.A. & PARMLEY, W.W. Passive smoking and heart disease: mechanisms and risk. *Journal of the American Medical Association*, **273**: 1047–1053 (1995).

7. WELLS, A.J. Passive smoking as a cause of heart disease. *Journal of the American College of Cardiology*, **24**: 546–554 (1994).
8. TAYLOR, A.E. ET AL. Environmental tobacco smoke and cardiovascular disease: a position paper from the Council on Cardiopulmonary and Critical Care, American Heart Association. *Circulation*, **86**: 1–4 (1992).
9. KRISTENSEN, T.S. Cardiovascular diseases and the work environment: a critical review of the epidemiologic literature on chemical factors. *Scandinavian journal of work, environment and health*, **15**: 245–264 (1989).
10. NATIONAL INSTITUTE FOR OCCUPATIONAL SAFETY AND HEALTH. *Posthearing Brief submitted to OSHA Docket No. H–122 (Submission #527)*. Washington, DC, Occupational Safety and Health Administration, 1995.
11. LAW, M.R. ET AL. Environmental tobacco smoke exposure and ischaemic heart disease: an evaluation of the evidence. *British medical journal*, **315**: 973–980 (1997).
12. CHILMONCZYK, B.A. ET AL. Association between exposure to environmental tobacco smoke and exacerbations of asthma in children. *New England journal of medicine*, **328**: 1665–1669 (1993).
13. STRACHAN, D.P. ET AL. Passive smoking, salivary cotinine concentrations, and middle ear effusion in 7-year-old children. *British medical journal*, **298**: 1549–1552 (1989).
14. RYLANDER, E. ET AL. Parental smoking, urinary cotinine, and wheezing bronchitis in children. *Epidemiology*, **6**: 289–293 (1995).
15. *The health benefits of smoking cessation. A report of the Surgeon General.* Washington, DC, US Department of Health and Human Services, 1990 (DHHS Publication No. (CDC) 90–8416).
16. TREDANIEL, J. ET AL. Environmental tobacco smoke and the risk of cancer in adults. *European journal of cancer*, **29A**: 2058–2068 (1993).
17. ZHENG, W. ET AL. Risk factors for cancers of the nasal cavity and paranasal sinuses among white men in the United States. *American journal of epidemiology*, **138**: 965–972 (1994).
18. HOFFMAN, H.J. & HILLMAN, L.S. Epidemiology of the sudden infant death syndrome: maternal, neonatal, and postnatal risk factors. *Clinics in perinatology*, **19**: 717–737 (1992).
19. SLOTKIN, T.A. ET AL. Loss of neonatal hypoxia tolerance after prenatal nicotine exposure: implications for sudden infant death syndrome. *Brain research bulletin*, **38**: 69–75 (1995).
20. MITCHELL, E.A. ET AL. Smoking and the sudden infant death syndrome. *Pediatrics*, **91**: 893–896 (1993).
21. KLONOFF-COHEN, H.S. ET AL. The effect of passive smoking and tobacco smoke exposure through breast milk on sudden infant death

syndrome. *Journal of the American Medical Association*, **273**: 795–798 (1995).
22. SCRAGG, R. ET AL. Bed sharing, smoking, and alcohol in the sudden infant death syndrome. *British medical journal*, **307**: 1312–1318 (1993).
23. WHO EUROPEAN CENTRE FOR ENVIRONMENT AND HEALTH. *Concern for Europe's tomorrow. Health and the environment in the WHO European Region.* Stuttgart, Wissenschaftliche Verlagsgesellschaft, 1995.

8.2 Man-made vitreous fibres

Exposure evaluation

Airborne concentrations during the installation of insulation comprising man-made vitreous fibres (MMVF) are in the range 10^5–2×10^6 fibres/m^3 *(1)*, which is generally higher than the concentrations of about 10^5 fibres/m^3 reported for production plants *(2)*. Little information is available on ambient concentrations of MMVF. A few limited studies of MMVF in outdoor air have reported concentrations ranging from 2 fibres/m^3 in a rural area to 1.7×10^3 fibres/m^3 near a city *(3–5)*. These levels are estimated to represent a very small percentage of the total fibre and total suspended particulate concentrations in the ambient air.

Health risk evaluation

MMVF of diameters greater than 3 µm can cause transient irritation and inflammation of the skin, eyes and upper airways *(6)*.

The deep lung penetration of various MMVF varies considerably, as a function of the nominal diameter of the material. For the six categories of MMVF considered here (continuous filament fibre glass, glass wool fibres, rock wool fibres, slag wool fibres, refractory ceramic fibres and special purpose fibres (glass microfibres)), the potential for deep lung penetration is greatest for refractory ceramic fibres and glass microfibres; both of these materials are primarily used in industrial applications.

In two large epidemiological studies, there have been excesses of lung cancer in rock/slag wool production workers, but not in glass wool, glass microfibre or continuous filament production workers. There have been no increases in the incidence of mesotheliomas in epidemiological studies of MMVF production workers *(7, 8)*. Although concomitant exposure to other substances may have contributed to the observed increase in lung cancer in the rock/slag wool production sector, available data are consistent with the hypothesis that the fibres themselves are the principal determinants of risk. Increases in tumour incidence have not been observed in inhalation studies in animals exposed to rock/slag wool, glass wool or glass microfibre, though they have occurred following intracavitary administration. Available data concerning the effects of continuous filament in animals are limited.

Several types of refractory ceramic fibre have been clearly demonstrated to be carcinogenic in inhalation studies in animal species, inducing dose-related increased incidence of pulmonary tumours and mesotheliomas in rats and hamsters *(9–11)*. Increased tumour incidence has also been observed following intratracheal *(12)* and intrapleural and intraperitoneal *(13)* administration in animals.

Though uses of refractory ceramic fibres are restricted primarily to the industrial environment, a unit cancer risk for lung tumours for refractory ceramic fibres has been calculated as 1×10^{-6} per fibre/l (for fibre length > 5 µm, and aspect ratio (ratio of fibre length to fibre diameter) of 3:1 as determined by optical microscopy) based on inhalation studies in animals *(14)*.

Guidelines

IARC classified rock wool, slag wool, glass wool and ceramic fibres in Group 2B (possibly carcinogenic to humans) while glass filaments were not considered classifiable as to their carcinogenicity to humans (Group 3) *(15)*. Recent data from inhalation studies in animals strengthen the evidence for the possible carcinogenicity of refractory ceramic fibres in humans.

Though uses of refractory ceramic fibres are restricted primarily to the industrial environment, the unit risk for lung tumours is 1×10^{-6} per fibre/l. The corresponding concentrations of refractory ceramic fibres producing excess lifetime risks of 1/10 000, 1/100 000 and 1/1 000 000 are 100, 10 and 1 fibre/l, respectively.

For most other MMVF, available data are considered inadequate to establish air quality guidelines.

References

1. ESMEN, N.A. ET AL. Exposure of employees to man-made mineral vitreous fibres: installation of insulation materials. *Environmental research*, **28**: 386–398 (1982).
2. DODGSON, J. ET AL. Estimates of past exposure to respirable man-made mineral fibres in the European insulation wool industry. *Annals of occupational hygiene*, **31**: 567–582 (1987).
3. BALZER, J.L. Environmental data: airborne concentrations found in various operations. *In: Occupational exposure to fibrous glass. Proceedings of a symposium, College Park, Maryland, June 26–27, 1974*. Washington, DC, US Department of Health, Education and Welfare, 1976.
4. HOHR, D. Transmissionselektronenmikroskopische Untersuchung: faserformige Staube in den Aussenluft [Investigation by means of transmission electron microscopy: fibrous particles in the ambient air]. *Staub Reinhaltung der Luft*, **45**: 171–174 (1985).

5. DOYLE, P. ET AL. *Mineral fibres (man-made vitreous fibres). Priority substances list assessment report.* Ottawa, Environment Canada and Health Canada, 1993.
6. *Man-made mineral fibres.* Geneva, World Health Organization, 1988 (Environmental Health Criteria, No. 77).
7. MARSH, G. ET AL. Mortality among a cohort of US man-made mineral fibre workers: 1985 follow-up. *Journal of occupational medicine*, **32**: 594–604 (1990).
8. SIMONATO, L. ET AL. The International Agency for Research on Cancer historical cohort study of MMMF production workers in seven European countries: extension of the follow up. *Annals of occupational hygiene*, **31**: 603–623 (1987).
9. DAVIS, J.M.G. ET AL. The pathogenicity of long versus short fibre sample of amosite asbestos administered to rats by inhalation and intraperitoneal injection. *British journal of experimental pathology*, **67**: 415–430 (1986).
10. MAST, R.W. ET AL. A multiple dose chronic inhalation toxicity study of kaolin refractory ceramic fiber (RFC) in male Fischer 344 rats. [Abstract no. 63]. *Toxicologist*, **13**: 43 (1993).
11. POTT, F. & ROLLER, M. Carcinogenicity of synthetic fibres in experimental animals: its significance for workers. *Journal of occupational health and safety – Australia and New Zealand*, **12**: 333–339 (1996).
12. POTT, F. ET AL. Significance of durability of mineral fibers for their toxicity and carcinogenic potency in the abdominal cavity of rats and the low sensitivity of inhalation studies. *Environmental health perspectives*, **102** (Suppl. 5): 145–150 (1994).
13. POTT, F. ET AL. Tumours by the intraperitoneal and interpleural routes and their significance for the classification of mineral fibres. *In*: Brown, R.C. et al., ed. *Mechanisms in fibre carcinogenesis.* New York & London, Plenum Press, 1991 (NATO ASI Series, Series A, Life Sciences, Vol. 223), pp. 547–565.
14. *Health-based tolerable intakes/concentrations and tumorigenic doses/concentrations for priority substances.* Ottawa, Health Canada, 1996.
15. *Man-made mineral fibres and radon.* Lyons, International Agency for Research on Cancer, 1988 (IARC Monographs on the Evaluation of Carcinogenic Risks to Humans, Vol. 43).

8.3 Radon

Exposure evaluation

Exposure to radon and radon progeny is the dominant source of exposure to ionizing radiation in most countries. The radon levels vary considerably between dwellings, and depend primarily on the inflow of soil gas and the type of building material. As shown in Table 28, arithmetic mean concentrations in European countries range from about 20 Bq/m^3 to 100 Bq/m^3, with even higher levels in some regions. The geometric mean concentrations are generally about 20–50% lower because of the skewed distribution of radon levels.

Health risk evaluation

A few recent case-control studies provide evidence on lung cancer risks related to residential radon exposure. In general, the exposure assessment was based on radon measurements in the homes of the people being studied, covering residential periods of about 10–30 years *(1)*. Some of the studies indicate increased relative risks for lung cancer by estimated time-weighted residential radon level or cumulative exposure, but the picture is not fully coherent. It should be realized, however, that most studies lacked an adequate statistical power. The largest of the studies, with analyses over the widest range of exposure, showed a clear increase in risk with estimated exposure to radon, which appeared consistent with a linear relative risk model *(2)*. The interaction between radon exposure and smoking with regard to lung cancer exceeded additivity and was close to a multiplicative effect.

To date, risk estimation for residential radon exposure has often been based on extrapolation of findings in underground miners. Several circumstances make such estimates uncertain for the general population, however, including the possible influence of other exposure factors in the mines and differences in age, sex, size distribution of aerosols, the attached fraction of radon progeny, breathing rate and route *(3, 4)*. Furthermore, the relevance is not fully understood of the apparent inverse effect of exposure rate observed in miners and the possible difference in relative risk estimates for nonsmokers and smokers *(5)*.

It is of interest to compare risk estimates based on the nationwide Swedish study on residential radon exposure and lung cancer *(2)* with those obtained from miners. Fig. 1 shows the estimated attributable proportion of lung

Table 28. Radon levels in dwellings of some European countries

Country	Number of houses sampled	Period and duration of exposure	Sample characteristics	Radon concentration (Bq/m^3)					
				Average	Geometric mean	Geometric mean SD[a]	Percentage over 200 Bq/m^3	Percentage over 400 Bq/m^3	Reference
Belgium	300	1984–1990 3 months to 1 year	population-based (selected acquaintances)	48	37	1.9	1.7	0.3	[b]
Czechoslovakia	1200	1982 random grab sampling	–	140	–	–	–	–	*(7)*
Denmark	496	1985–1986 6 months	random	47	29	2.2	2.2	< 0.4	*(8)*
Finland	3074	1990–1991 1 year	random	123	84	2.1	12.3	3.6	*(9)*
France	1548	1982–1991 3 months (using open alpha track detectors)	biased (not stratified)	85	52	2.3	7.1	2.3	*(10)*
Germany	7500	1978–1984 3 months 1991–1993 1 year	random	50	40	–	1.5–2.5	0.5–1	*(11, 12)*

Greece	73	1988 6 months	–	52	–	–	–	–	*(7)*
Hungary	122	1985–1987 2.5 years	preliminary survey	55	42 (median)	–	–	–	[c]
Ireland	1259	1985–1989 6 months	random	60	34	2.5	3.8	1.6	*(13)*
Italy	4866	1989–1994 1 year	stratified random	75	62	2.0	4.8	1.0	*(14)*
Luxembourg	2500	1991	–	–	65	–	–	–	*(7)*
Netherlands	1000	1982–1984 1 year	random	29	24 (median)	1.6	–	–	*(7, 15)*
Norway	7525	1987–1989 6 months	random	60	32	–	5.0	1.6	*(16)*
Portugal	4200	1989–1990 1–3 months	volunteers in a selected group (high school students)	81	37	–	8.6	2.6	*(17)*
Spain	1555–2000	winter of 1988–1989 grab sampling	random	86	41–43	2.6–3.7	–	4	*(7, 18)*
Sweden	1360	1982–1992 3 months in heating season	random	108	56	–	14	4.8	*(19)*

Table 28. (contd)

Country	Number of houses sampled	Period and duration of exposure	Sample characteristics	Radon concentration (Bq/m^3)					Reference
				Average	Geometric mean	Geometric mean SD[a]	Percentage over 200 Bq/m^3	Percentage over 400 Bq/m^3	
Switzerland	1540	1982–1990 3 months (mainly in winter)	biased (not stratified)	70	–	–	5.0	–	*(20)*
United Kingdom	2093	1986–1987 1 year	random	20.5	15	2.2	0.5	0.2	*(21)*

[a] SD = Standard deviation.

[b] A. Poffjin, personal communication.

[c] L. Sztanyik & I. Nikl, personal communication.

Source: Bochicchio et al. *(22)*.

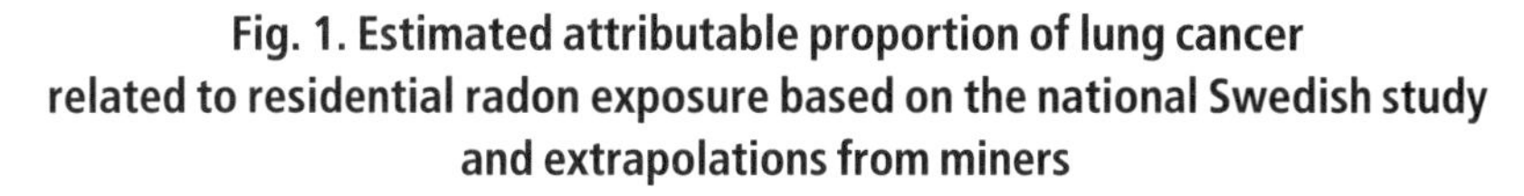

Fig. 1. Estimated attributable proportion of lung cancer related to residential radon exposure based on the national Swedish study and extrapolations from miners

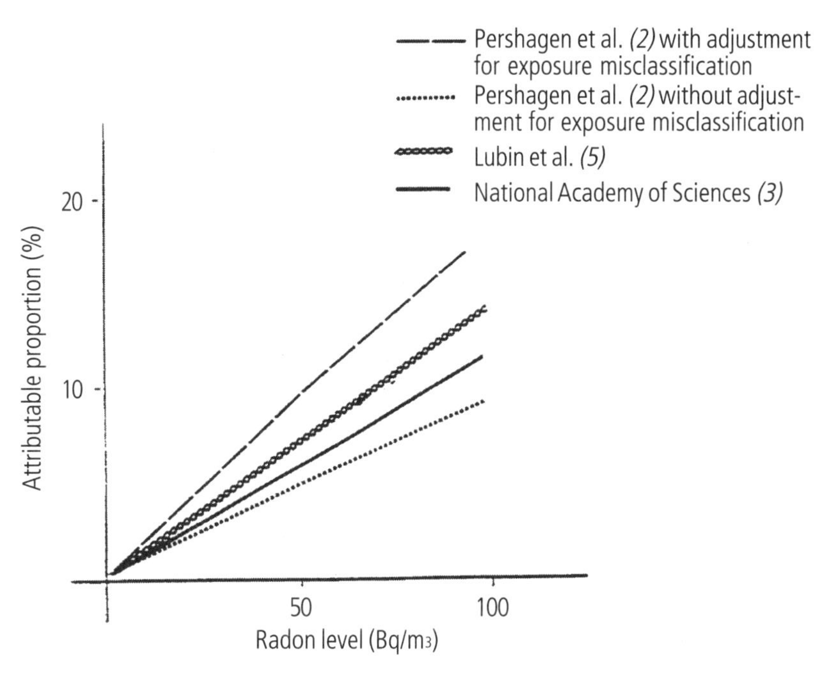

cancer related to residential radon, using risk estimates from the Swedish study and assuming a linear relative risk model. Imprecision in the exposure estimation leads to attenuation of the exposure–response relationship, and it has been indicated that this may have led to an underestimation of the risk by a factor of up to about 2 *(6)*. It is suggested that the true values lie between the unadjusted and adjusted estimates.

Fig. 1 also gives estimates of attributable proportion based on extrapolations from underground miners, after adjusting for dosimetric differences between mines and homes. As an example, the radon concentration distribution in western Germany, with an arithmetic mean of 50 Bq/m^3, leads to an attributable proportion of 7% (95% confidence interval: 1–29%) using the model in Lubin et al. *(5)*, and 6% (95% confidence interval: 2–17%) using that of the National Academy of Sciences *(3)*. Corresponding values based on the Swedish residential study are 5% and 9%, respectively, without and with adjustment for exposure misclassification.

Table 29 shows population risk estimates under three different assumptions with regard to population exposure, taken to represent long-term residential exposure in European countries with relatively high, medium and low residential radon concentrations. The estimated attributable proportion of lung cancer related to residential radon exposure ranges from 2–5% in low-exposure areas to 9–17% in high-exposure areas.

Table 29 also shows estimated excess lifetime deaths from lung cancer related to residential radon. Assuming that lung cancer deaths constitute 3% of total deaths, it is estimated that around 600–1500 excess lung cancer deaths occur per million people exposed on average to

Table 29. Attributable proportion of lung cancer related to long-term residential radon exposure in regions with high, medium and low indoor concentrations [a]

	Concentration		
	High	**Medium**	**Low**
Radon concentration			
Arithmetic mean (Bq/m^3)	100	50	25
> 200 Bq/m^3	15%	1.5%	0.75%
> 400 Bq/m^3	5%	0.5%	0.25%
Proportion of all lung cancers attributable to the exposure			
Total	9–17% [b]	5–9%	2–5%
> 200 Bq/m^3	4–6%	0.4–0.6%	0.2–0.3%
> 400 Bq/m^3	2–3%	0.2–0.3%	0.1–0.15%
Excess lifetime lung cancer deaths (per million) [c]			
Total	2700–5100	1500–2700	600–1500
> 200 Bq/m^3	1200–1800	120–180	60–90
> 400 Bq/m^3	600–900	60–90	30–45

[a] A linear relative risk model is assumed and a multiplicative interaction between radon and other risk factors for lung cancer, including smoking.

[b] The range in estimated attributable proportion is based on assessment of the uncertainty due to imprecision in exposure estimates of the observed exposure–response relationship *(6)*.

[c] It is assumed that lung cancer deaths constitute 3% of total deaths.

Source: Pershagen et al. *(2)*.

25 Bq/m^3 over their lifetime. For an average exposure of 100 Bq/m^3, the corresponding estimate ranges from 2700 to 5100 excess lung cancer deaths per million people exposed.

Guidelines

Radon is a known human carcinogen (classified by IARC as Group 1 *(23)*) with genotoxic action. No safe level of exposure can be determined. Quantitative risk estimates may be obtained from a recent large residential study, which are in general agreement with a linear extrapolation of risks observed in miners. The risk estimates obtained in the studies conducted among miners and the recent study from Sweden *(2)* would correspond to a unit risk of approximately 3–6 × 10^{-5} per Bq/m^3, assuming a lifetime risk of lung cancer of 3%. This means that a person living in an average European house with 50 Bq/m^3 has a lifetime excess lung cancer risk of 1.5–3 × 10^{-3}. Similarly, a person living in a house with a high radon concentration of 1000 Bq/m^3 has a lifetime excess lung cancer risk of 30–60 × 10^{-3} (3–6%), implying a doubling of background lung cancer risk.

Current levels of radon in dwellings and other buildings are of public health concern. A lifetime lung cancer risk below about 1 × 10^{-4} cannot be expected to be achievable because natural concentration of radon in ambient outdoor air is about 10 Bq/m^3. No guideline value for radon concentration is recommended. Nevertheless, the risk can be reduced effectively based on procedures that include optimization and evaluation of available control techniques. In general, simple remedial measures should be considered for buildings with radon progeny concentrations of more than 100 Bq/m^3 equilibrium equivalent radon as an annual average, with a view to reducing such concentrations wherever possible.

References

1. *Indoor air quality – a risk-based approach to health criteria for radon indoors*. Copenhagen, WHO Regional Office for Europe, 1996 (document EUR/ICP/CEH 108(A)).
2. PERSHAGEN, G. ET AL. Residential radon exposure and lung cancer in Sweden. *New England journal of medicine*, **330**: 159–164 (1994).
3. NATIONAL ACADEMY OF SCIENCES. *Health risks of radon and other internally deposited alpha-emitters*. Washington, DC, National Academy Press, 1988.
4. NATIONAL RESEARCH COUNCIL. *Comparative dosimetry of radon in mines and houses*. Washington, DC, National Academy Press, 1991.

5. LUBIN, J.H. ET AL. *Radon and lung cancer – a joint analysis of 11 underground miners studies.* Washington DC, National Cancer Institute, 1994 (NIH publication No. 94-3644).
6. LAGARDE, F. ET AL. Residential radon and lung cancer in Sweden: risk analysis accounting for random error in the exposure assessment. *Health physics*, **72**: 269–276 (1997).
7. UNITED NATIONS SCIENTIFIC COMMITTEE ON THE EFFECTS OF ATOMIC RADIATION. *Report to the General Assembly, with scientific annexes.* New York, United Nations, 1993.
8. ULBAK, K. ET AL. Results from the Danish indoor radiation survey. *Radiation protection dosimetry*, **24**: 402–405 (1988).
9. ARVELA, H. ET AL. *Otantatutkimus asuntojen radonista Suomessa* [Radon in a sample of Finnish houses]. Helsinki, Finnish Centre for Radiation and Nuclear Safety, 1993.
10. RANNOU, A. ET AL. *Campagnes de mesure de l'irradiation naturelle gamma et radon en France. Bilan de 1977 a 1990.* Rapport SEGR No. 10, 1992.
11. URBAN, M. ET AL. *Bestimmung der Strahlenbelastung der Bevölkerung durch Radon und dessen kurzlebige Zerfallsprodukte in Wohnhäuser und im Freien.* Karlsruhe, Kernforschungszentrum, 1985.
12. CZARWINSKI, R. ET AL. Investigations of the radon concentrations in buildings of Eastern Germany. *Annals of the Association of Belgian Radioprotection*, **19**: 175–188 (1994).
13. MCLAUGHLIN, J.P. & WASIOLEK, P. Radon levels in Irish dwellings. *Radiation protection dosimetry*, **24**: 383–386 (1988).
14. BOCHICCHIO F. ET AL. Results of the representative Italian national survey on radon indoors. *Health physics,* **71**: 743–750 (1996).
15. PUT, L.W. ET AL. Survey of radon concentrations in Dutch dwellings. *Science of the total environment*, **45**: 441–448 (1985).
16. STRAND, T. ET AL. Radon in Norwegian dwellings. *Radiation protection dosimetry*, **45**: 503–508 (1992).
17. FAÍSCA, M.C. ET AL. Indoor radon concentrations in Portugal – a national survey. *Radiation protection dosimetry*, **45**: 465–467 (1992).
18. QUINDOS, L.S. et al. National survey of indoor radon in Spain. *Environment international*, **17**: 449–453 (1991).
19. SWEDJEMARK, G.A. ET AL. Radon levels in the 1988 Swedish housing stock. *In*: Proceedings of Indoor Air '93, Vol. 4, pp. 491–496. Helsinki, Indoor Air '93, 1993.
20. SURBECK, H. & VÖLKLE, H. Radon in Switzerland. *In*: *Proceedings of International Symposium on Radon and Radon Reduction Technology, Philadelphia, 1991, Vol 3, paper VI–3.* Research Triangle Park, NC, US Environmental Protection Agency, 1991.

21. WRIXON, A.D. ET AL. *Natural radiation exposure in UK dwellings. NRPB–R190.* Oxford, National Radiological Protection Board, 1988.
22. BOCHICCHIO F. ET AL. *Radon in indoor air. Report No. 15, European Collaborative Action: indoor air quality and its impact on man.* Luxembourg, Office for Official Publications of the European Communities, 1995.
23. *Man-made mineral fibres and radon.* Lyons, International Agency for Research on Cancer, 1988 (IARC Monographs on the Evaluation of Carcinogenic Risks to Humans, Vol. 43).

PART III

EVALUATION OF ECOTOXIC EFFECTS

CHAPTER 9

General approach

In the context of the updating and revision of these guidelines, the ecological effects of major air pollutants were considered in more detail. This was undertaken in cooperation with the Working Group on Effects under the United Nations Economic Commission for Europe (ECE) Convention on Long-range Transboundary Air Pollution, capitalizing on the scientific work undertaken since 1988 to formulate criteria for the assessment of the effects of air pollutants on the natural environment.

The evaluation for the guidelines focused on the ecological effects of sulfur dioxide (including sulfur and total acid deposition), nitrogen dioxide (and other nitrogen compounds including ammonia) and ozone, which were thought to be currently of greatest concern across Europe. A number of other atmospheric contaminants are known to have ecological effects, but were not considered by the working groups. In the case of metals and persistent organic pollutants, levels of soil contamination or bioaccumulation leading to adverse effects have been proposed, but methods of linking these to atmospheric concentrations or depositions have not yet been developed. In the case of fluorides and particles, ecological effects are no longer of widespread concern in Europe, although air quality criteria have been proposed in the past by other bodies, and new criteria for fluorides are currently under consideration by certain national governments.

USE OF THE GUIDELINES IN PROTECTING THE ENVIRONMENT

Although the main objective of the guidelines is the direct protection of human health, the WHO strategy for health for all recognizes the importance of protecting the environment in terms of benefits to human health and wellbeing. Resolution WHA42.26 of the World Health Assembly and resolutions 42/187 and 42/186 of the United Nations General Assembly recognize the interdependence of health and the environment.

Ecologically based guidelines for preventing adverse effects on terrestrial vegetation were included for the first time in the first edition of *Air quality guidelines for Europe* in 1987, and guidelines were recommended for some gaseous air pollutants. Since that time, however, significant advances have been made in the scientific understanding of the impacts of air pollutants on the environment. The realization that soils play an important role in mediating both the direct and indirect effects of air pollutants on terrestrial and freshwater ecosystems has led to the development and acceptance of the joint concepts of critical levels and critical loads within the framework of the ECE Convention on Long-range Transboundary Air Pollution.

At the ECE Workshop on Critical Loads for Sulphur and Nitrogen, held at Skokloster, Sweden *(1)* and at a workshop on critical levels held at Bad Harzburg, Germany *(2)*, the following definitions were agreed on.

Critical level is the concentration of pollutants in the atmosphere above which direct adverse effects on receptors such as plants, ecosystems or materials may occur according to present knowledge.

Critical load is a quantitative estimate of an exposure, in the form of deposition, to one or more pollutants below which significant harmful effects on specified sensitive elements of the environment do not occur according to present knowledge.

The critical levels and loads approach is essentially a further development of the first edition of these guidelines published in 1987. There are several fundamental differences between conventional environmental objectives, critical levels and critical loads (Table 30).

Critical levels relate to direct effects on plant physiology, growth and vitality, and are expressed as atmospheric concentrations or cumulative exposures over a given averaging time. Typically, critical levels are based on effects observed over periods of from one day to several years. Critical loads relate to effects on ecosystem structure and functioning, and are expressed as annual depositions of mass or acidity. Typically, critical loads relate to the potential effects over periods of decades. In the case of sulfur and nitrogen compounds, critical levels can be directly related to critical loads when the deposition velocity for a given vegetation type is known. Nevertheless, while critical levels provide effects thresholds for relatively short-term exposures, and are not aimed at providing complete protection of all plants in all situations from adverse effects, critical loads provide the long-term deposition below which we are sure that adverse ecosystem effects will not occur.

Both critical levels and critical loads may be used to indicate the state of existing or required environmental protection, and they have been used by

Table 30. Differences between conventional environmental objectives, critical levels and critical loads

Conventional objectives	Critical levels	Critical loads
Effects are generally experienced at the organism level	Effects are experienced from organism to ecosystem levels	Effects are usually manifested at the ecosystem level
Objectives are established on the basis of laboratory tests	Objectives are established by laboratory or controlled environmental and field studies	Ecosystem studies are required to establish values
Lethality or physiological effects are the usual response used in setting objectives	Physiological, growth and ecosystem effects are caused by direct or indirect mechanisms	Ecosystem effects are caused by direct (abiotic change) or indirect (biotic interaction) mechanisms
Environmental objectives are set well below known effects to provide some margin of safety	Objectives are set as close to effect thresholds as possible	Objectives are set as close to effect thresholds as possible
No beneficial effects are likely to occur in the environment at any level	Changes may occur that are deemed beneficial (such as increased growth)	Changes may occur that are deemed beneficial (such as increased productivity)
Environmental damage from exceedances is usually observed within a short time	Environmental damage usually results from short- to medium-term exceedances	Environmental damage usually results from long-term (years, decades) exceedances and may be cumulative

ECE to define air pollutant emission control strategies for the whole of Europe. They are being or may be used in a series of protocols relating to the control of sulfur dioxide, nitrogen oxides, total nitrogen (including oxidized and reduced species) and ozone. Full use has been made in this publication of the data that underpin these protocols. The proposed guidelines cover the same range of air pollutants and are aimed at a wide range of vegetation types and ecosystems. Individual species, vegetation types and ecosystems may vary in their sensitivity to a given pollutant, and this sensitivity may also depend on other factors such as soil type or climate. When possible, therefore, different values of critical loads or levels are defined, depending on the relevant factors. When this approach is not possible, values are based on protecting the most sensitive type of vegetation or ecosystem for which good quality data are available.

There is thus a sound scientific basis for expecting that adverse ecological and economic effects may occur when the guidelines recommended below are exceeded. There is a possibility that adverse effects might also occur at exposures below these guidelines, but there is considerable uncertainty over this and it was decided to recommend values with a sound scientific basis rather than to incorporate arbitrary uncertainty factors. Critical levels and critical loads thus fulfil the primary aim of air quality guidelines in providing the best available sound scientific basis for the protection of vegetation from significant effects.

To carry out an assessment based on the guidelines, due consideration has to be given to the various problems caused by air pollution and their impact on the stock that may be at risk. The requirements for the former are often different from those needed to assess the risks to human health. Nevertheless, methodologies have been developed that can assess the risks of damage to vegetation and ecosystems.

Because of the different definition of critical loads and critical levels, the variable nature of the ecological impacts caused by different pollutants, and the different types of scientific evidence available, it is not possible to use a single methodology to derive the air quality guidelines presented in this section. For critical levels, the methods used rely on analysis either of experimental studies in the laboratory or in field chambers, or of field studies along pollution gradients. For critical loads, the methods used rely on analysis of field experiments, comparisons of sites with different deposition rates, or modelling. Where possible, data from a combination of sources are used to provide the strongest support for the proposed guidelines. Uncertainties in defining guidelines can arise (*a*) because of the limited availability of appropriate data; (*b*) because the data exist only for specific vegetation types and climates and therefore may not be representative of all areas of Europe; or (*c*) because exposure patterns in experimental chambers may not be representative of those under field conditions.

In the field, pollutants are never present in isolation, while the same pollutant may have several impacts simultaneously (for example, exposure to sulfur dioxide can cause direct effects on leaf physiology and contribute to long-term acidification, while deposition of nitrogen can cause both acidification and eutrophication). Currently, knowledge of the impacts of pollutant combinations is inadequate to define critical loads or levels for such combined impacts, and thus the guidelines are recommended for the ecological effects of individual pollutants. When applying these guidelines in ecological risk assessment, the possibility of such combined impacts should be considered. Furthermore, when considering an area of mixed vegetation types or ecosystems, several guidelines may apply. Thus ecological risk assessment applying the critical levels and loads approach must be aimed at identifying or protecting the most sensitive element of the environment.

A simple overview of the elements of how critical levels and critical loads can be used is given in Fig. 2. The left- and right-hand pathways indicate the requirements, enabling finally the comparison of critical levels or critical loads with ambient air concentrations (present levels) or pollutant depositions (present loads) on broad spatial scales. The left-hand pathway depicts the steps needed to obtain a geographical distribution of critical levels and loads over European ecosystems.

Since critical levels and critical loads indicate the sensitivity of receptors (such as individual plant species or ecosystems) to air pollutants, an important step in the critical levels/loads application pathway consists of the geographical determination and mapping of the receptors and their sensitivities, at as fine a spatial resolution as possible.

Critical levels are in most cases formulated in such a way that a certain receptor type (such as forests or crops) has the same critical level value throughout Europe. In these cases, the resulting sensitivity maps look uniform over large areas. More recent developments in critical levels research attempt to incorporate environmental conditions into the assessment. The incorporation of such modifying factors – such as water availability, which influences the opening of the stomata and thus the uptake of gaseous pollutants by plants – can lead to a higher degree of differentiation in the mapping of sensitivities.

Critical loads are also allocated to certain receptor types, such as forests, bogs, heathlands, grasslands or lakes, but the spatial differentiation is generally more advanced than in the case of critical levels. It is often possible to take into account environmental conditions such as soil characteristics, water conditions, precipitation amounts, land use and management practices. The result is a critical load map with a high spatial variation in sensitivities.

Fig. 2. Critical levels/loads application pathway

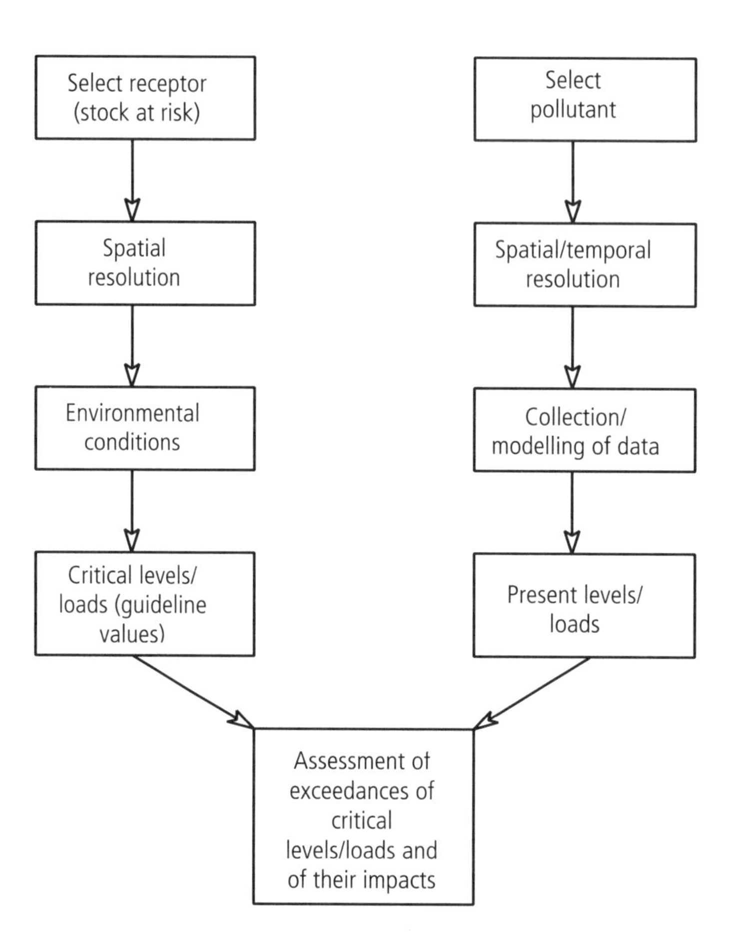

The right-hand pathway in Fig. 2 depicts steps to ensure comparability of present levels/loads with critical levels/loads. The comparison with present ambient air concentrations or present depositions can only be made if the spatial resolution is compatible with the mapped critical levels/loads. The regional distribution of ambient air concentrations and depositions can be modelled to reflect data measured by national and/or international monitoring networks over Europe. Subject to the spatial resolution of these modelled data, comparisons of critical levels/loads with present levels/loads can be made at finer or coarser spatial resolutions. At the European level, present levels/loads are currently modelled for grid cells with a size of 150 km × 150 km or 50 km × 50 km by the ECE Co-operative

Programme for Monitoring and Evaluation of the Long Range Transmission of Air Pollutants in Europe (EMEP). In the case of depositions, compatibility with the mapped critical loads can be achieved by establishing cumulative frequency distributions of the critical loads occurring in the grid cell. A low percentile value (such as 5) of these distributions can be chosen for comparison with the present loads. If, in the framework of effect-orientated pollutant emission reduction strategies, the present levels or loads are reduced to critical levels or a 5-percentile value of the critical loads distribution, respectively, the protection of most sensitive receptors is reliably estimated to be high (for example achieving potential protection of 95% of the ecosystems in a grid cell).

The left- and right-hand pathways of Fig. 2 finally lead to the assessment of exceedances of critical levels/loads. Exceedances of critical levels/loads are interpreted in a qualitative rather than a quantitative manner, in that the probability of damage is considered to be non-zero whenever critical levels/loads are exceeded. Thus, the exceedance of critical levels/loads implies non-sustainable stress, which can lead to damage at any point in time and to an extent depending on the amount of excess pollution. Research is continuing to determine quantitative regional relationships between the actual excess pollution and the expected damage. Exposure–response relationships for sensitive receptors, established in experimental or field studies and modified for prevailing environmental conditions, may tentatively be used to quantify the consequences of excess pollution. However, research results are considered to lack the robustness needed to allow applications to European ecosystems as a whole.

REFERENCES

1. Nilsson, J. & Grennfelt, P., ed. *Critical loads for sulphur and nitrogen.* Copenhagen, Nordic Council of Ministers, 1993 (Miljørapport No. 15).
2. *Final report of the Critical Levels Workshop, Bad Harzburg, Germany, 14–18 March 1988.* Berlin, Federal Environment Agency, 1988.

CHAPTER 10

Effects of sulfur dioxide on vegetation: critical levels

Since the publication of the first edition of the *Air quality guidelines for Europe* in 1987 *(1)*, the relative importance of sulfur dioxide as a phytotoxic pollutant in Europe has diminished to some extent, owing to falling emissions in many areas. In terms of understanding the basic mechanisms of direct injury by sulfur dioxide, and threshold concentrations for adverse effects, advances have been made in demonstrating the significance of very low concentrations on growth and yield and on changing plant sensitivity to other environmental stresses. New work has also provided information that can be utilized to introduce new guidelines for protecting lichens against sulfur dioxide and forests against acid mists.

A number of studies have provided valuable data for several major agricultural crops, based on fumigations, filtrations and transect studies *(2–4)*. These new data confirm the annual guideline value of 30 μg/m^3 as an annual mean concentration (Table 31). However, it is recommended that this value should also not be exceeded as a mean concentration for the winter months (October–March inclusive) in view of the abundant evidence for increased sensitivity of crops growing slowly under winter conditions. It is recommended that the 24-hour air quality guideline for all species be abandoned, in view of further evidence confirming that peak concentrations are not significant compared with the accumulated dose.

A lower air quality guideline of 20 μg/m^3 is now recommended for forests and natural vegetation, as both an annual and winter mean concentration (Table 31). This is based on new evidence of periods of high sensitivity of conifers during needle elongation and the longevity of many of the species concerned as well as their being unmanaged or minimally managed, which renders them more sensitive to pollution stress *(4–6)*.

New data have confirmed concerns over low-temperature stress contributing to greater sulfur dioxide sensitivity in forests. Further justification for modifying the air quality guideline to take account of interactions with low temperature is given by evidence of sulfate mists enhancing frost sensitivity.

Table 31. Guidelines for the effects of sulfur dioxide on vegetation: critical levels

Vegetation category	Guideline ($\mu g/m^3$)	Time period [a]	Constraints
Agricultural crops	30	Annual and winter mean	
Forests and natural vegetation	20	Annual and winter mean	
Forests and natural vegetation	15	Annual and winter mean	Accumulated temperature sum above +5 °C is < 1000 °C·days per year
Lichens	10	Annual mean	
Forests	1.0 sulfate particulate[b]	Annual mean	Where ground level cloud is present ≥ 10% of time

[a] Where annual and winter mean concentrations are indicated, the higher value should be used to define exceedance. Winter is defined as October to March inclusive.
[b] Air quality guideline only applies in areas of oceanic Europe where calcium and magnesium concentrations in cloud or mist are less than the combined ionic concentrations of H^+ and NH_4^+.

A field study of Norway spruce at different altitudes in the Ore Mountains of Czechoslovakia has been used to develop a model, from which the accumulated temperature sum above +5 °C of < 1000 °C·days per year is used as a threshold for lowering mean annual and winter sulfur dioxide concentrations to 15 μg/m³ for protecting forests and natural vegetation. This lower concentration is now recommended as a WHO air quality guideline for regions below this threshold temperature sum (Table 31). It should be recognized, however, that this guideline is based on field studies in a region where the temperatures recorded were above those pertaining in some areas of northern Europe, and thus it is possible that in even more extreme environments a lower guideline is required.

The 1987 edition of the guidelines considered only the effects of sulfur dioxide on higher plants. Many sensitive lichen and bryophyte species have disappeared from large areas of Europe with only moderately elevated sulfur dioxide concentrations. Annual mean concentrations of 30 μg/m³ are associated with the eradication of the most sensitive lichen taxa. On the basis of new field studies, it is recommended that an air quality guideline of 10 μg/m^3 annual mean (Table 31) be established for lichens *(7–9)*.

In the 1987 edition, no consideration was given to direct impacts of acid precipitation on above-ground plant organs. It is now recognized that mists can contain solute concentrations up to ten times those of rain, and can thus have a direct impact on vegetation. Since mists and clouds occur most frequently at high altitudes, and are intercepted with particular efficiency by forests, trees are likely to be the most sensitive receptors. Experiments on young trees, backed up by field observations, show significant effects of acid mists on leaf surface structure at pH 3.5, which is equivalent to 150 μmol/l sulfate. Because of the difficulties of measuring sulfate concentrations in cloud water, a guideline has been set based on the equivalent particulate sulfate concentration. A guideline of 1.0 μg/m^3 particulate sulfate as an annual mean is recommended for trees where ground level cloud is present 10% or more of the time (Table 31). This guideline only applies, however, when calcium and magnesium concentrations in cloud do not exceed hydrogen and ammonium ion concentrations, because no data exist to establish a guideline under other conditions. This restriction excludes areas such as the Mediterranean region, eastern Europe and the Alps.

These guidelines do not take into account that sulfur dioxide increases sensitivity to other stresses, with the exception of low temperatures for forests and natural vegetation. Given further knowledge of its effects on stresses such as drought, pathogens and pests, it is possible that the guidelines may require further modification in the future. The 24-hour mean guideline has been abolished, but this is on the basis of knowledge on higher plants. The inclusion of lichens in these new guidelines may warrant future considerations of a short-term guideline for these organisms, if knowledge indicates the necessity for this. The new guideline for acid mists has similarly been set for forests only, and the effects on other receptors may also warrant future attention.

REFERENCES

1. *Air quality guidelines for Europe*. Copenhagen, WHO Regional Office for Europe, 1987 (WHO Regional Publications, European Series, No. 23).
2. JÄGER, H.J. ET AL., ED. *Effects of air pollution on agricultural crops in Europe*. Brussels, European Commission, 1993 (Air Pollution Research Report, No. 46).

3. KRUPA, S.G. & ARNDT, U. Special issue on the Hohenheim Long Term Experiment. *Environmental pollution*, **68**: 193–478 (1990).
4. MCLEOD, A. R. & SKEFFINGTON, R. A. The Liphook Forest Fumigation Project – an overview. *Plant, cell and environment*, **18**: 327–336 (1995).
5. SCHULZE, E.D. ET AL. Forest decline and air pollution; a study of spruce *(Picea abies)* on acid soils. Springer-Verlag, 1989 (Ecological Studies, No. 77).
6. MCLEOD, A. R. ET AL. Enhancement of nitrogen deposition to forest trees exposed to SO_2. *Nature*, **347**: 272–279 (1990).
7. FIELDS, R.F. Physiological responses of lichens to air pollutant fumigations. *In*: Nash, T.H. & Wirth, V., ed. *Lichens, bryophytes and air quality*. Berlin, Cramer, 1988, pp. 175–200.
8. RICHARDSON, D.H.A. Understanding the pollution sensitivity of lichens. *Botanical journal of the Linnean Society*, **96**: 31–43 (1988).
9. WINNER, W.E. Responses of bryophytes to air pollution. *In*: Nash, T.H. & Wirth, V., ed. *Lichens, bryophytes and air quality*. Berlin, Cramer, 1988, pp. 141–173.

Effects of nitrogen-containing air pollutants: critical levels

EFFECT EVALUATION

Various forms of nitrogen pollute the air, mainly nitric oxide (NO), nitrogen dioxide (NO_2) and ammonia (NH_3) as dry deposition, and nitrate (NO_3^-) and ammonium (NH_4^+) as wet deposition. Other contributions are from occult deposition (fog, clouds, aerosols), peroxyacetylnitrate (PAN), dinitrogen pentoxide (N_2O_5), nitrous oxide (N_2O) and amines. Since the publication of the *Air quality guidelines for Europe* in 1987 *(1)* there have been significant advances in knowledge of the impacts of nitrogen oxides (NO_x, i.e. NO_2 and NO) and NH_3 on vegetation.

In the present evaluation, attention is mainly paid to direct effects on plants caused by an exposure duration of between one hour and one year. The long-term impact (more than one year) on vegetation and the nitrogen cycle is discussed in Chapter 14, while the contribution of nitrogen-containing air pollutants to soil acidification is evaluated in Chapter 13. The properties of PAN are discussed in Chapter 12. The role of NO_x and N_2O in atmospheric chemistry (formation and depletion of ozone in the troposphere and stratosphere, respectively) and relations with climate change are not considered.

The reason for defining critical levels for NO, NO_2 and NH_3 is the recent evidence from monitoring and mapping that these are the dominant forms of nitrogen deposition in many parts of the world, and that several important effects of these compounds are not covered by the critical loads for nitrogen or acidity.

The critical levels are based on a survey of published evidence of physiological and ecologically important effects on plants *(2–7)*. Biochemical changes have only been used as additional indicators of potentially relevant ecological responses. The current survey has considered that, in an ecological context, growth stimulation and reduction are both potentially negative responses. For instance, both NO_x and NH_y (i.e. NH_3 and NH_4^+) generally cause an increase in the shoot:root ratio, which may or may not be beneficial.

Responses to nitrogenous pollutants can be further modified and exacerbated by interactions with other environmental factors, including frost, drought and pest organisms. These interactions generally include increased susceptibility to these factors, which may in turn lead to major ecological changes.

The method of estimating critical levels is different for NO_x and NH_3, but both are based on a 95% protection level (neglecting the 5% lowest effective exposures).

GAPS IN KNOWLEDGE

There have been important developments in the use of critical level and critical load approaches for setting air quality guidelines. With regard to the critical levels of nitrogen-containing air pollutants, however, there are several areas where improvements are urgently required.

- The guidelines for the critical levels of NO_x and NH_3 are intended to apply to all classes of vegetation and under all environmental conditions. However, more information is needed to quantify the range of sensitivity.

- The guideline for NH_3 is based on research performed in temperate climates on a limited range of soil types. To a lesser extent this applies to NO_x as well. Caution is required when critical levels are considered for plants in very different conditions, for example in tropical and subtropical zones.

- There is a need to understand further the long-term impacts on growth of changes in biochemical parameters.

- There is growing awareness of the physiological importance of NO, and this is reflected in the new incorporation of this compound in the guideline for NO_x. Comparisons of the phytotoxicity of NO and NO_2 are scarce and still not conclusive with regard to their relative degree of toxicity.

- The relevance of the emission of NH_3 from plants should be investigated in more detail in order to establish its potential importance in nitrogen budgets.

GUIDELINES

Evidence exists that NH_4^+ (and NO_3^-) in rain, clouds and fog can have significant direct effects on vegetation, but current knowledge is still insufficient to arrive at critical levels for those compounds. It is assumed that NO and NO_2 act in an additive manner.

A strong case can be made for the provision of critical levels for short-term exposures. There are insufficient data to provide these levels with confidence at present, but current evidence suggests values of about 75 $\mu g/m^3$ for NO_x and 270 $\mu g/m^3$ for NH_3 as 24-hour means.

Interactive effects between NO_2 and sulfur dioxide and/or ozone have been reported frequently *(8–13)*. From a review of recent literature, however, it was concluded that the lowest effective levels for NO_2 are approximately equal to those for combination effects (although in general, at concentrations near to its effect threshold, NO_2 causes growth stimulation if it is the only pollutant, while in combination with sulfur dioxide and/or ozone it results in growth inhibition).

Critical levels for a 1-year period are recommended to cover relatively long-term effects. The critical level for NO_x (NO and NO_2, added in ppb and expressed as NO_2 in $\mu g/m^3$) is 30 $\mu g/m^3$ as an annual mean. The critical level for NH_3 is 8 $\mu g/m^3$ as an annual mean.

REFERENCES

1. *Air quality guidelines for Europe*. Copenhagen, WHO Regional Office for Europe, 1987 (WHO Regional Publications, European Series, No. 23).
2. ASHENDEN, T.W. ET AL. *Critical loads of N & S deposition to semi-natural vegetation*. Bangor, Institute for Terrestrial Ecology, 1990 (Report Project T07064L5).
3. BOBBINK, R. ET AL. Critical loads for nitrogen eutrophication of terrestrial and wetland ecosystems based upon changes in vegetation and fauna. *In*: Grennfelt, P. & Thörnelöf, E., ed. *Critical loads for nitrogen*. Copenhagen, Nordic Council of Ministers, 1992, pp. 111–159.
4. FANGMEIJER, A. ET AL. Effects of atmospheric ammonia on vegetation – a review. *Environmental pollution*, **86**: 43–82 (1994).
5. GRENNFELT, P. & THÖRNELÖF, E., ED. *Critical loads for nitrogen*. Copenhagen, Nordic Council of Ministers, 1993 (Miljørapport No. 41).
6. GUDERIAN, R. *Critical levels for effects of NO_x*. Geneva, United Nations Economic Commission for Europe, 1988.

7. SCHULZE, E.D. ET AL. *Forest decline and air pollution; a study of spruce (*Picea abies*) on acid soils*. Heidelberg, Springer-Verlag, 1989 (Ecological Studies, No. 77).
8. ADAROS, G. ET AL. Concurrent exposure to SO_2 alters the growth and yield responses of wheat and barley to low concentrations of CO_2. *New phytologist*, **118**: 581–591 (1991).
9. ADAROS, G. ET AL. Single and interactive effects of low levels of O_3, SO_2 and NO_2 on the growth and yield of spring rape. *Environmental pollution*, 72: 269–286 (1991).
10. CAPE, J.N. ET AL. Sulfate and ammonium in mist impair the frost hardening of red spruce seedlings. *New phytologist*, **118**: 119–126 (1991).
11. CAPORN, T.M. ET AL. Canopy photosynthesis of CO_2-enriched lettuce (*Lactuca sativa* L.). Response to short term changes in CO_2, temperature and oxides of nitrogen. *New phytologist*, **126**: 45–52 (1994).
12. ITO, O. ET AL. *Effects of NO_2 and O_3 alone or in combination on kidney bean plants. II. Amino acid pool size and composition*. Tokyo, National Institute of Environmental Studies, 1984 (Research Report No. 66).
13. VAN DE GEIJN, S.C. ET AL. Problems and approaches to integrating the concurrent impacts of elevated CO_2, temperature, UVb radiation and O_3 on crop production. *In*: Buxton, D.R. et al., ed. Madison, WI, Crop Science of America, 1993, pp. 333–338.

CHAPTER 12

Effects of ozone on vegetation: critical levels

EFFECT EVALUATION

The revision of the air quality guidelines for ozone builds on the progress made to define critical levels to protect crops and tree species. Guidelines for other photochemical oxidants, such as peroxyacetylnitrate and hydrogen peroxide, are not recommended because of the low levels of these pollutants observed in Europe, and because data concerning their effects on plants in Europe are very limited. Research in recent years has mainly advanced our understanding of the exposure, uptake and effects of ozone *(1, 2)*.

Ozone concentrations vary widely both in space and in time, and in order to quantitatively relate ozone exposure to effects it is necessary to summarize the concentration pattern during the exposure period in a biologically meaningful way *(3, 4)*. From results of exposure–response studies with open-top chambers, it is concluded that mean concentrations are not appropriate to characterize ozone exposure. This is mainly because (*a*) the effect of ozone results from the cumulative exposure; and (*b*) not all concentrations are equally effective, higher concentrations having greater effects than lower concentrations. Ozone exposure is therefore expressed as the sum of all 1-hour mean concentrations above a cut-off concentration of 40 ppb. It is emphasized that 40 ppb should not be regarded as a lower concentration limit for biological effects, since some biological effects may occur below this value; rather, it is a cut-off concentration used to calculate an exposure index that is strongly related to biological responses, and hence to the degree of risk to sensitive vegetation.

The use of 40 ppb as the cut-off concentration provides good linear relationships between ozone exposure and plant response for a number of species, thus confirming its biological relevance *(5, 6)*. Furthermore, the ozone concentrations found in most areas of Europe, in the absence of photochemical pollution, are in the range 10–40 ppb, except at very high altitudes. In relation to long-term effects, this sum (referred to as the "Accumulated exposure Over a Threshold of 40 ppb", AOT40), is calculated for a 3-month growing season in the case of crops or herbaceous semi-natural vegetation, or a 6-month growing season for trees. The appropriate months to define the growing season will depend on the vegetation and climate in a specific region or at a specific site. Since uptake of ozone by vegetation occurs primarily during daylight hours when stomata are open, the calculation of the AOT40 considers only those hours when radiation is higher than 50 W/m^2.

To define critical levels, the AOT40 is related to specific effects *(2)*. A reduction in economic yield (such as grain yield in wheat) is considered the most relevant long-term effect of ozone on crop species, and a reduction in biomass is chosen for tree species. For semi-natural vegetation, the effect of ozone is expressed as the change in the species composition. The most important short-term effect of ozone is the appearance of visible leaf injury. The most sensitive species for each vegetation type for which adequate data are available was selected to derive the critical level.

For crops, data on grain yield of spring wheat exposed in open-top field chambers to different ozone concentrations over the growing season were used to set the critical level, since the database is the largest (10 experiments in 6 countries using 10 different cultivars) and most consistent, and wheat is known to be a sensitive species. Statistical analysis of this pooled dataset showed that the least significant deviation in yield that can be estimated with 99% confidence is 4–5%. The critical level determined using this criterion (Table 32) is 3 ppm·h *(5, 7)*.

The critical level for short-term effects of ozone on crops (visible injury) is derived from an extensive database of coordinated European field observations, involving eight countries over two growing seasons using two clover species *(8)*. Using artificial neural network analysis, combinations of ozone exposure and climatic conditions in the five days preceding the onset of visible injury were identified and used to set critical levels (Table 32) of 0.2 ppm·h for humid air conditions (mean vapour pressure deficit below 1.5 kPa) and 0.5 ppm·h for dry air conditions (mean vapour pressure deficit above 1.5 kPa).

For forests, the database available is small. Data sets from three different European studies using open-top field chambers of the effects of ozone on annual biomass increment in beech saplings have been used *(9)*. Statistical analysis of these data showed that the least significant deviation in biomass increment that could be estimated with 95% confidence was about 10%, and this criterion was used to determine a critical level of 10 ppm·h (Table 32).

Table 32. Guidelines for the effects of ozone on vegetation: critical levels

Vegetation type	Guidelines AOT40 (ppm·h)	Time period[a]	Constraints
Crops (yield)	3	3 months	
Crops (visible injury)	0.2	5 days	Humid air conditions (mean daytime VPD[b] below 1.5 kPa)
	0.5	5 days	Dry air conditions (mean daytime VPD[b] above 1.5 kPa)
Forests	10	6 months	
Semi-natural vegetation	3	3 months	

[a] Daylight hours.

[b] VPD = vapour pressure deficit.

Finally, for herbaceous species of semi-natural vegetation, recent studies have reported the effects of ozone in field or laboratory chambers on shoot biomass, seed biomass or relative growth rate of a total of 87 species. All studies showed significant adverse effects, at the 95% confidence level, of exposures in the range 3–5 ppm·h on the most sensitive species studied. Since there is also evidence that the most sensitive of these species are as sensitive as the most sensitive known crop species, a critical level of 3 ppm·h (Table 32), equivalent to that for crops, has been adopted *(10)*.

GUIDELINES

The data used to derive critical levels are almost entirely drawn from experiments in open-top chambers in central and northern Europe, using plants that are adequately supplied with water and nutrients. There are uncertainties in using these data to define air quality guidelines for vegetation throughout Europe. Among the most important of these uncertainties are the following.

- The open-top chamber technique will tend to overestimate the effects because of the higher ozone fluxes within the chambers compared with outside.
- There are a great many species that have not been investigated experimentally in Europe, especially in the Mediterranean region.

- The critical level is likely to be higher when water availability is limited, because ozone flux is reduced. This is a very significant factor in many areas of Europe, especially as periods of water stress often coincide with periods of high ozone concentration.
- There may be physiological, morphological or biochemical changes induced by ozone exposures below the critical level that could be important, for example in altering sensitivity to other abiotic and biotic stresses.
- The data on trees are more variable than those for annual crops and there is uncertainty about the extent and significance of night-time ozone uptake. Furthermore, there are uncertainties in extrapolating from experiments of limited duration with young pot-grown trees to long-term effects on forest ecosystems. For these reasons, there is greater uncertainty attached to the recommended guidelines for trees.
- For changes in species composition, the experiments are also of limited duration, and there is great uncertainty about the long-term effects of ozone exposure.

When determining whether ozone exposures at a specified location exceed the critical levels (Table 32), two points need to be carefully considered.

1. Over short vegetation, but not over forests, there may be significant gradients in AOT40 immediately above the vegetation, and thus AOT40 values determined at the measurement height of most monitoring stations may be larger than at the surface of the vegetation. In contrast, in experimental chambers used to generate the exposure–response data, the air is well mixed and the gradients do not exist.

2. AOT40 values can vary substantially from year to year, because of the variability of the climate. Because the critical level for crop yield was based on analysis of data in several different growing seasons, and because the critical level for forests was based on multi-year experiments, it is recommended that the exceedance of these critical levels, and that for semi-natural vegetation, be evaluated on the basis of mean AOT40 values over a 5-year period. Where visible injury to crops resulting from short-term exposures is of direct economic concern, however, examination of monitoring data for the year with highest ozone exposures is recommended.

REFERENCES

1. JÄGER, H.J. ET AL., ED. *Effects of air pollution on agricultural crops in Europe*. Brussels, European Commission, 1993 (Air Pollution Research Report, No. 46).

2. FUHRER, J. & ACHERMANN, B., ED. *Critical levels for ozone: a UN-ECE workshop report.* Liebefeld-Bern, Swiss Federal Research Station for Agricultural Chemistry and Environmental Hygiene, 1994 (FAC Report, No. 16).
3. DERWENT, R.G. & KAY, P.J.A. Factors influencing the ground level distribution of ozone in Europe. *Environmental pollution*, **55**: 191–219 (1988).
4. RUNECKLES, V.C. Dosage of air pollutants and damage to vegetation. *Environmental conservation*, **1**: 305–308 (1974).
5. ASHMORE, M.R. & WILSON, R.B., ED. *Critical levels for air pollutants in Europe.* London, Department of the Environment, 1994.
6. FUHRER, J. The critical level for ozone to protect agricultural crops. An assessment of data from European open-top chamber experiments. *In*: Fuhrer, J. & Achermann, B., ed. *Critical levels for ozone: a UN-ECE workshop report.* Liebefeld-Bern, Swiss Federal Research Station for Agricultural Chemistry and Environmental Hygiene, 1994 (FAC Report, No. 16), pp. 42–57.
7. SKÄRBY, L. ET AL. Responses of cereals exposed to air pollutants in open-top chambers. *In*: Jäger, H.J. et al., ed. *Effects of air pollution on agricultural crops in Europe.* Brussels, European Commission, 1993 (Air Pollution Research Report, No. 46), pp. 241–259.
8. BENTON, J. ET AL. The critical level of ozone for visible injury on crops and natural vegetation (ICP Crops). *In*: Kärenlampi, L. & Skärby, L., ed. *Critical levels for ozone in Europe: testing and finalizing the concepts. UN-ECE workshop report.* Kuopio, Department of Ecology and Environmental Science, University of Kuopio, 1996, pp. 44–57.
9. BRAUN, S. & FLÜCKIGER, W. Effects of ambient ozone on seedlings of *Fagus silvatica* L. and *Picea abies* (L.) Karst. *New phytologist*, **129**: 33–44 (1995).
10. ASHMORE, M.R. & DAVISON, A.W. Toward a critical level of ozone for natural vegetation. *In*: Kärenlampi, L. & Skärby, L., ed. *Critical levels for ozone in Europe: testing and finalizing the concepts. UN-ECE workshop report.* Kuopio, Department of Ecology and Environmental Science, University of Kuopio, 1996, pp. 58–71.

CHAPTER 13

Indirect effects of acidifying compounds on natural systems: critical loads

ACIDIFYING DEPOSITION AND ECOSYSTEM DAMAGE

Historical data provide evidence of increasing transport of sulfate (SO_4^{2-}), up to a factor of 2 and 3.5 in 1950 and 1980, respectively, compared to pre-industrial levels in Europe. Emissions of sulfur dioxide in the air are transformed to sulfate, which constitutes the major compound of acid deposition *(1, 2)*. The effects and risks of sulfur dioxide emissions and resulting deposition are described for soils in general and for forest soils and surface waters in particular.

Soil acidification is defined as a decrease in the acid neutralizing capacity (ANC) of the inorganic fraction of the soil including the solution phase, and is directly dependent on the net supply of base cations (by weathering and deposition) and the net supply of anions (deposition minus retention) in the mineral soil *(3, 4)*. Deposition of acidifying compounds such as sulfur dioxide, nitrogen oxides and ammonia leads to soil acidification by oxidation to sulfuric and nitric acids and leaching of sulfate and nitrate, respectively.

The dynamics of forest soil acidification is very site-specific and depends on soil characteristics such as weathering rate, sulfate adsorption capacity and cation exchange capacity. The acidification of soils ultimately leads to an increase in the soil solution of the aluminium concentration, which increases the risk of vegetation damage. By defining the relationship between the chemical status (base cation and aluminium concentrations in the soil solution) and vegetation response, the so-called critical load for that particular ecosystem can be derived *(5)*. Damage to forests in Europe, including defoliation, discoloration, growth decrease and tree dieback, have been reported over the last decade, and have to a large extent been attributed to soil acidification, but also to eutrophication and photochemical oxidant effects.

Acidic deposition has caused acidification of surface waters, fish mortality and other ecological changes in large areas of northern Europe and eastern parts of North America.

Sulfate is normally a mobile anion in catchments located in glaciated areas. Increased sulfate concentrations in runoff due to increased acidifying inputs are accompanied by an increase of base cations and a decrease in bicarbonates, resulting in an acidifying effect on surface waters.

For most of the sensitive soils in Europe, the sulfate deposition is directly related to the acidifying load of sulfate in watershed runoff. The deposition/runoff relationship for nitrogen is not as well defined. The nitrate concentration in runoff, and hence the contribution to the total acidity loading, is due to a combination of factors including the amount of deposition, the ability of vegetation to take up nitrogen and the denitrifying processes. Even in cases of substantial nitrogen runoff, models calculate a deposition/runoff ratio greater than 1 due to denitrification. Determining the nitrate runoff response to a change in nitrogen deposition requires site-specific information.

Under natural conditions, most of the nitrogen deposited on terrestrial catchments is taken up by vegetation, leading to low concentrations of ammonia and nitrate in the runoff. In some areas in Europe, however, including Denmark, southern Norway and southern Sweden, nitrogen concentrations in runoff water appear to be above background values. This excess nitrate in runoff is mostly due to a disruption of the nitrogen cycle and not only to increased nitrogen deposition. In such cases, nitrogen deposition exceeds the rate of nitrogen retention mechanisms, i.e. growth uptake, denitrification and immobilization. When nitrate is leached from the soil solution and appears in surface waters, it will contribute to soil and surface water acidification in the same manner as sulfate.

In cases of low soil pH, excess nitrogen deposition leads to acidification of natural vegetation systems other than trees. Plant species from poorly buffered habitats are adapted to nitrate uptake, while plants from acid environments are generally adapted to ammonia uptake. A low pH may thus lead to a shift of these systems from a nitrate-dominated to an ammonia-dominated system. Such a disruption of the nitrogen cycle in combination with low pH ultimately leads to acidification by nitrogen.

CRITICAL LOADS

The critical load of acidity means "the highest deposition of compounds that will not cause chemical changes leading to harmful effects on ecosystem structure and function" *(6)*.

A relationship has been established between increased aluminium concentrations in the soil solution and adverse effects to roots and growth of trees. For example, it has been shown that the tree growth of Norway spruce decreases as the base cation (calcium, magnesium, potassium) to aluminium (BC/Al) ratio is smaller than a critical limit of 1 *(7)*. Other critical limits for forest soils are based on aluminium concentration and pH in soil solution. Laboratory results of aluminium damage indicate that tolerance to aluminium varies among tree species. For example, a growth reduction of 80% has been demonstrated at a BC/Al ratio of 0.1 for the northern white cedar (*Thuja occidentalis*) and of 4 for the masson pine (*Pinus massoniana*). It has been found that a BC/Al ratio exceeding or equal to 1 seems to provide appropriate sustainability for European forests. However, species that grow in non-glaciated old soils rich in aluminium oxide, such as teak, guapira, orange and cotton, seem to be more accustomed to aluminium than trees from the temperate zone. Computations of critical loads in Europe have therefore generally applied a BC/Al ratio of 1 *(7)*.

For surface waters, the ANC has been considered a chemical criterion that is used to explain the increased risk of damage to fish populations. The critical chemical value, ANC limit = 20 μeq/l, has been derived from the information on water chemistry and fish status obtained from the 1000-lake survey carried out in Norway in 1986 *(8, 9)*. The selected ANC limit was assessed by examining the relationship between the critical load exceedance, and damage to fish populations on the basis of data from the Norwegian 1000-lake survey can again be used *(10)*. The probability of damage to fish populations increases clearly as a function of the critical load exceedance. Table 33 gives an overview of average limits that have been established to compute critical loads.

Calculation of critical loads is based on the steady state mass balance method which assumes a time-independent steady state of chemical interaction involving an equilibrium between the production and the consumption of acidic compounds.

Current United Nations Economic Commission for Europe (ECE) protocols concentrate on distinctive acidifying compounds, such as sulfur and nitrogen, rather than on acidity as a whole. It was necessary to subdivide

Table 33. Critical limits for chemical compounds and properties in forest soils and freshwater systems

Compound / property	Unit	Forest soil	Fresh water	Groundwater
Aluminium	mol_c/m^3	0.2	0.003	0.02
BC/Al	mol/mol	1	–	–
pH	–	4.0 [a]	(5.3, 6.0) [b]	6.0
ANC	mol_c/m^3	–	(0.02, 0.08) [b]	0.14
NO_3^-	mol_c/m^3	–	–	0.8

[a] Assuming log K_{gibb} of 8.0 and Al = 0.2 mol_c/m^3.
[b] A pH of 6.0 relates to peak flow situations and is associated with ANC = 0.08 mol_c/m^3.

the critical load of acidity between the acidifying share of sulfur and that of nitrogen. For the purposes of the guidelines no subdivisions are performed.

GUIDELINES

In Europe, critical loads have been established at the EMEP resolution (see page 225) to allow for comparisons between critical loads and sulfur deposition values, and to identify areas where critical loads are exceeded. Critical loads of acidity, as computed by the steady state mass balance method, depend predominantly on the rate of base cation weathering. For terrestrial ecosystems, the weathering rate can be estimated by combining information on soil parent material and texture properties. The critical loads of acidity in relation to combinations of parent material and texture classes range from smaller than 250 eq/ha per year to more than 1500 eq/ha per year (see Table 34).

Additional factors, such as vegetation cover, further modify the value of the critical load. To calculate precise critical loads for a given geographical area, it is recommended that the mass balance equation be used. For surface waters, the weathering rate can be estimated on the basis of water quality and quantity variables, of which base cation concentrations and runoff are the most influential ones.

Table 34. Critical load ranges of acidity used for the various combinations of parent material and texture in terrestrial ecosystems

Guideline range of critical loads of acidity (eq/ha per year)	Parent material[a]	Texture[b]
<250	acidic	coarse
250–500	acidic	coarse-medium
	intermediate	coarse
	basic	coarse
500–1000	acidic	medium, medium-fine
	intermediate	coarse-medium, medium
	basic	coarse-medium
1000–1500	intermediate	medium-fine
	basic	medium
>1500	intermediate	fine
	basic	medium-fine

[a] Acidic: sand (stone), gravel, granite, quartzine, gneiss (schist, shale, greywacke, glacial till).
Intermediate: gronodiorite, loess, fluvial and marine sediment (schist, shale, greywacke, glacial till).
Basic: gabbro, basalt, dolomite, volcanic deposits.

[b] Coarse: clay content < 18%.
Medium: clay content 18–35%.
Fine: clay content > 35%.

Table 35 lists the ranges of critical loads in relation to combinations of base cation concentration and runoff classes. For each critical load class, at least 50% of the critical load values computed on the basis of lake data from Finland (1450 lakes), Norway and Sweden fall within the class boundaries, given the ranges for present base cation concentrations and runoff. Only in two cases did the boundaries for the base cation concentration classes overlap between two critical load classes, when the class boundaries were set on the basis of the 25th and 75th percentile base cation concentrations for given runoff classes. For those cases the critical loads are determined more by other factors than base cation levels and runoff, and the guideline value set is therefore more uncertain than those without overlap.

Table 35. Critical load ranges of acidity used for various combinations of base cation concentration and runoff for surface waters

Guideline range of critical loads of acidity (eq/ha per year)	Base cation concentration (meq/m^3)	Runoff (m)
<250	<45	>1.0
	<100	0.3–1.0
	<270 [a]	<0.3
250–500	45–70	>1.0
	100–190	0.3–1.0
	250–400 [a]	<0.3
500–1000	70–103	>1.0
	190–290	0.3–1.0
	400–650	<0.3
1000–1500	103–170	>1.0
	290–465 [a]	0.3–1.0
	650–1300	<0.3
>1500	>170	>1.0
	>350 [a]	0.3–1.0
	>1300	<0.3

[a] The class boundaries overlap.

REFERENCES

1. FINKEL, R.C. ET AL. Changes in precipitation chemistry at Dye 3, Greenland. *Journal of geophysics research*, **910**: 9849–9855 (1986).
2. TUOVINEN, J.-P. ET AL. *Transboundary acidifying pollution in Europe: calculated fields and budgets 1985–93*. Oslo, Norwegian Meteorological Institute, 1994 (Technical Report, No. 129).
3. VAN BREEMEN, N. ET AL. Acidic deposition and internal proton sources in acidification of soils and waters. *Nature*, **307**: 599 (1984).
4. DE VRIES, W. & BREEUWSMA, A. The relation between soil acidification and element cycling, *Water, air and soil pollution*, **35**: 293–310 (1987).

5. De Vries, W. *Soil response to acid deposition at different regional scales: field and laboratory data, critical loads and model predictions*. Dissertation, Agricultural University, Wageningen, 1994.
6. Nilsson, J. & Grennfelt, P., ed. *Critical loads for sulphur and nitrogen.* Copenhagen, Nordic Council of Ministers, 1993 (Miljørapport No. 15).
7. Sverdrup, H. & Warfvinge, P. *The effect of soil acidification on the growth of trees, grass and herbs as expressed by the (Ca+Mg+K)/Al ratio.* Lund, University of Lund, 1993 (Report No. 2).
8. Henriksen, A. et al. Lake acidification in Norway. Present and predicted chemical status. *Ambio,* **17**: 259–266 (1988).
9. Henriksen, A. et al. Lake acidification in Norway. Present and predicted fish status. *Ambio,* **18**: 314–321 (1989).
10. Henriksen, A. & Hesthagen, T. *Critical load exceedance and damage to fish populations*. Oslo, Norwegian Institute for Water Research, 1993 (Project Naturens Tålegrenser, Fagrapport No. 43).

CHAPTER 14

Effects of airborne nitrogen pollutants on vegetation: critical loads

Most of earth's biodiversity is found in natural and seminatural ecosystems, both in aquatic and terrestrial habitats. Man's activities pose a number of threats to the structure and functioning of these ecosystems, and thus to the natural variety of plant and animal species. One of the major threats in recent years is the increase in airborne nitrogen pollution, namely NH_y (consisting of ammonia and ammonium ions), and NO_x (consisting of nitrogen dioxide and nitric oxide). Nitrogen is the limiting nutrient for plant growth in many of these ecosystems. Most of the plant species from these habitats are adapted to nutrient-poor conditions, and can only compete successfully on soils with low nitrogen levels *(1)*. Nitrogen is the only nutrient whose cycle through the ecosystem is almost exclusively regulated by biological processes.

To establish reliable critical loads for nitrogen, it is essential to understand the effects of nitrogen on these ecosystem processes. The critical loads for nitrogen depend on:

- the type of ecosystem;
- the land use and management in the past and present; and
- the abiotic conditions, especially those that influence the nitrification potential and immobilization rate in the soil.

The impacts of increased nitrogen deposition on biological systems are diverse, but the most important effects are:

- short-term direct effects of nitrogen gases and aerosols on individual species (see Chapter 11);
- soil-mediated effects;
- increased susceptibility to secondary stress factors; and
- changes in (competitive) relationships between species, resulting in loss of biodiversity.

The empirical approach has been used to establish guidelines for excess nitrogen deposition on natural and seminatural vegetation. It was decided not to include the results of the mass balance approach with nitrogen as a nutrient for non-forest ecosystems, because essential data are missing. The acidifying effects of airborne nitrogen are incorporated in the guidelines for excess acidity based on steady state mass balance models (see Chapter 13).

EVALUATION OF CRITICAL LOADS

The main aim of this evaluation was to update the guideline for airborne nitrogen deposition on vegetation, which was estimated at 30 kg/ha per year for sensitive vegetation *(2)*. Since 1987, significant progress has been made in understanding the ecological effects of nitrogen deposition on several types of vegetation. Critical loads of nitrogen have been formulated on an empirical basis by observing changes in the vegetation, fauna and biodiversity *(3, 4)*. Experiments under controlled and field conditions, and comparisons of vegetation and fauna composition in time and space, are used to detect changes in ecosystem structure *(5–7)*.

Changes in plant development and in species composition or dominance have been used as a "detectable change" for the impacts of excess nitrogen deposition, but in some cases a change in ecosystem function, such as nitrogen leaching or nitrogen accumulation, has been used. The results of dynamic ecosystem models, integrating both biotic and abiotic processes, are also used where available. Based on these data, guidelines for nitrogen deposition (critical loads) have been presented for receptor groups of natural and seminatural ecosystems, namely:

- wetlands, bogs and softwater lakes
- species-rich grasslands
- heathlands
- forest ecosystems (including tree health and biodiversity).

Critical loads have been defined within a range per ecosystem, because of (*a*) real intra-ecosystem variation within and between countries, (*b*) the range of experimental treatment where an effect was observed or not observed, or (*c*) uncertainties in deposition values, where critical loads are based on field observations. The reliability of the figures presented is shown in Table 36.

It is advised, where insufficient national data are available, to use the lower, middle or upper part of the ranges of the nitrogen critical loads for terrestrial

Table 36. Guidelines for nitrogen deposition to natural and seminatural freshwater and terrestrial ecosystems

Ecosystem	Critical load [a] (kg N/ha per year)	Indication of exceedance
Wetlands		
Softwater lakes	5–10 ##	Decline in isoetid aquatic plant species
Ombrotrophic (raised) bogs	5–10 #	Decrease in typical mosses; increase in tall graminoids; nitrogen accumulation
Mesotrophic fens	20–35 #	Increase in tall graminoids; decline in diversity
Species-rich grasslands		
Calcareous grasslands	15–35 #	Increase in tall grasses; decline in diversity [b]
Neutral–acid grasslands	20–30 #	Increase in tall grasses; decline in diversity
Montane–subalpine grassland	10–15 (#)	Increase in tall graminoids; decline in diversity
Heathlands		
Lowland dry heathland	15–20 ##	Transition from heather to grass
Lowland wet heathland	17–22 #	Transition from heather to grass
Species-rich heaths/acid grassland	10–15 #	Decline in sensitive species
Upland *Calluna* heaths	10–20 (#)	Decline in heather, mosses and lichens
Arctic and alpine heaths	5–15 (#)	Decline in lichens, mosses and evergreen dwarf shrubs; increase in grasses
Trees and forest ecosystems		
Coniferous trees (acidic; low nitrification rate)	10–15 ##	Nutrient imbalance
Coniferous trees (acidic; moderate–high nitrification rate)	20–30 #	Nutrient imbalance

Table 36. (contd)

Ecosystem	Critical load [a] (kg N/ha per year)	Indication of exceedance
Trees and forest ecosystems (contd)		
Deciduous trees	15–20 #	Nutrient imbalance; increased shoot/root ratio
Acidic coniferous forests	7–20 ##	Changes in ground flora and mycorrhizas; increased leaching
Acidic deciduous forests	10–20 #	Changes in ground flora
Calcareous forests	15–20 (#)	Changes in ground flora
Acidic forests (unmanaged)	7–15 (#)	Changes in ground flora and leaching
Forests in humid climates	5–10 (#)	Decline in lichens; increase in free-living algae

[a] ## *Reliable*: a number of published papers on various types of study show comparable results.
Fairly reliable: the results of some studies are comparable.
(#) *Expert judgement*: no data are available for this type of ecosystem; the critical load is based on knowledge of ecosystems likely to be more or less comparable with this ecosystem.

[b] Use low end of the range for nitrogen-limited and high end for phosphorus-limited calcareous grasslands.

receptor groups according to the general relationships between abiotic factors and critical loads for nitrogen (Table 37).

At this moment, the critical loads are set in values of total nitrogen inputs. More information is needed in future on the relative effects of oxidized and reduced nitrogen deposition. Critical loads for nitrogen are formulated as reliably as possible. As most research has focused on acidification in forestry, serious gaps in knowledge exist on the effects of enhanced nitrogen deposition on natural and seminatural terrestrial and aquatic ecosystems. The following gaps in knowledge are particularly important:

- more research is needed in Mediterranean, tropical and subtropical vegetation zones;
- quantified effects of enhanced nitrogen deposition on fauna in all types of vegetation reviewed are extremely scarce;
- the critical loads for nitrogen deposition to Arctic and alpine heathlands and forests are largely speculative;

- more research is needed on the effects of nitrogen on forest ground vegetation and (ground) fauna, because most research had focused on the trees only;
- there is a serious gap in knowledge on the effects of nitrogen on neutral/calcareous forests that are not sensitive to acidification;
- more long-term research is needed in montane/subalpine meadows, species-rich grasslands and ombrotrophic bogs;
- the long-term effects of enhanced atmospheric nitrogen in grassland and heathland of great importance for nature conservation under different management regimes are insufficiently known and may affect the critical load value;
- the possible differential effects of the deposited nitrogen species are insufficiently known for the establishment of critical loads; and
- the long-term effects of nitrogen eutrophication in (sensitive) aquatic ecosystems (freshwater and marine) need further research.

GUIDELINES

To establish reliable guidelines, it is crucial to understand the long-term effects of increased nitrogen deposition on ecological processes in a representative range of ecosystems. It is thus very important to quantify the effects of nitrogen loads on natural and seminatural terrestrial and freshwater ecosystems by manipulation of nitrogen inputs in long-term ecosystem studies in unaffected and affected areas. These data are essential to validate the presented critical loads and to develop robust dynamic ecosystem models reliable enough to calculate critical loads for nitrogen deposition in such ecosystems.

Table 37. Suggestions for using the lower, middle or upper part of the set critical loads of terrestrial ecosystems (excluding wetlands) if national data are insufficient

Action	Temperature	Soil wetness	Frost period	Base cation availability
Move to lower part	Cold	Dry	Long	Low
Use middle part	Intermediate	Normal	Short	Intermediate
Move to higher part	Hot	Wet	None	High

Guidelines for nitrogen deposition to natural and seminatural ecosystems are given in Table 36. The most sensitive ecosystems have critical loads of 5–10 kg N/ha per year. An average value for natural and seminatural ecosystems is 15–20 kg N/ha per year.

REFERENCES

1. TAMM, C.O. *Nitrogen in terrestrial ecosystems. Questions of productivity, vegetational changes, and ecosystem stability*. Berlin, Springer-Verlag, 1991.
2. The effects of nitrogen on vegetation. *In*: *Air quality guidelines for Europe*. Copenhagen, WHO Regional Office for Europe, 1987 (WHO Regional Publications, European Series, No. 23), pp. 373–385.
3. BOBBINK, R. ET AL. Critical loads for nitrogen eutrophication of terrestrial and wetland ecosystems based upon changes in vegetation and fauna. *In*: Grennfelt, P. & Thörnelöf, E., ed. *Critical loads for nitrogen*. Copenhagen, Nordic Council of Ministers, 1992, pp. 111–159.
4. ROSEN, K. ET AL. Nitrogen enrichment of Nordic forest ecosystems – the concept of critical loads. *Ambio*, **21**: 364–368 (1992).
5. HENRIKSEN, A. Critical loads of nitrogen to surface water *In*: Nilsson, J. & Grennfelt, P., ed *Critical loads for sulphur and nitrogen*. Copenhagen, Nordic Council of Ministers, 1993 (Miljørapport No. 15), pp. 385–412.
6. HULTBERG, H. Critical loads for sulphur to lakes and streams. *In*: Nilsson, J. & Grennfelt, P., ed *Critical loads for sulphur and nitrogen*. Copenhagen, Nordic Council of Ministers, 1993 (Miljørapport No. 15), pp.185–200.
7. KÄMÄRI, J. ET AL. Nitrogen critical loads and their exceedance for surface waters. *In*: Grennfelt, P. & Thörnelöf, E., ed. *Critical loads for nitrogen*. Copenhagen, Nordic Council of Ministers, 1993 (Nord Miljörapport No. 41), pp. 161–200.

Participants at WHO air quality guideline meetings

PLANNING MEETING ON THE UPDATE AND REVISION OF THE AIR QUALITY GUIDELINES FOR EUROPE BILTHOVEN, NETHERLANDS, 11–13 JANUARY 1993

Temporary Advisers

Dr Ole-Anders Braathen
Head, Organic Chemistry Laboratory, Norwegian Institute for Air Research, Lillestrøm, Norway

Dr Lester D. Grant
Environmental Criteria and Assessment Office, US Environmental Protection Agency, Research Triangle Park, NC, USA

Dr Robert L. Maynard
Senior Medical Officer, Department of Health, London, United Kingdom *(Rapporteur)*

Dr Peter J.A. Rombout
Laboratory for Toxicology, National Institute of Public Health and Environmental Protection, Bilthoven, Netherlands

Professor Bernd Seifert
Institut für Wasser, Boden und Lufthygiene, Berlin, Germany *(Chairperson)*

Professor Hans-Urs Wanner
Institut für Hygiene und Arbeitsphysiologie, Zurich, Switzerland

Representatives of other organizations

European Commission

Mr Pierre Hecq
Environment, Nuclear Safety and Civil Protection, Brussels, Belgium

World Health Organization

Regional Office for Europe

Dr Michal Krzyzanowski
Epidemiologist, WHO European Centre for Environment and Health, Bilthoven, Netherlands

Ms Barbara Lübkert
Air Pollution Scientist, WHO European Centre for Environment and Health, Bilthoven, Netherlands

Dr Maged Younes
Toxicologist, WHO European Centre for Environment and Health, Bilthoven, Netherlands *(Scientific Secretary)*

Headquarters

Dr Bingheng Chen
Toxicologist, International Programme on Chemical Safety, Geneva, Switzerland

Dr David Mage
Prevention of Environmental Pollution, Geneva, Switzerland

WORKING GROUP ON METHODOLOGY AND FORMAT FOR UPDATING AND REVISING THE AIR QUALITY GUIDELINES FOR EUROPE BILTHOVEN, NETHERLANDS, 20–22 SEPTEMBER 1993

Temporary Advisers

Dr Bert Brunekreef
Department of Epidemiology and Public Health, Agricultural University, Wageningen, Netherlands

Dr Judith A. Graham
Associate Director, Environmental Criteria and Assessment Office, US Environmental Protection Agency, Research Triangle Park, NC, USA

Professor Marek Jakubowski
Scientific Secretary, Nofer's Institute of Occupational Medicine, Lodz, Poland

Professor Paul J. Lioy
Director, Exposure Measurement and Assessment Division, Environmental and Occupational Health Sciences Institute, Piscataway, NJ, USA

Dr Robert L. Maynard
Head, Air Pollution Section, Department of Health, London, United Kingdom *(Rapporteur)*

Dr Peter J.A. Rombout
Department of Toxicology, National Institute of Public Health and Environmental Protection, Bilthoven, Netherlands *(Chairperson)*

Professor Bernd Seifert
Director, Institute for Water, Soil and Air Hygiene of the Federal Health Office, Berlin, Germany

Dr Per E. Schwarze
Head, Section of Air Pollution Toxicology, Department of Environmental Medicine, National Institute of Public Health, Oslo, Norway

Dr Katarina Victorin
Toxicologist, Institute of Environmental Medicine, Karolinska Institute, Stockholm, Sweden

Professor Giovanni A. Zapponi
Director, Environmental Impact Assessment Unit, Istituto Superiore di Sanità, Rome, Italy

Representatives of other organizations

European Commission

Ms Kathleen Cameron
Detached National Expert, DG XI-B3, Brussels, Belgium

World Health Organization

Regional Office for Europe

Dr Michal Krzyzanowski
Epidemiologist, WHO European Centre for Environment and Health, Bilthoven, Netherlands

Dr Maged Younes
Toxicologist, WHO European Centre for Environment and Health, Bilthoven, Netherlands

Headquarters

Dr Bingheng Chen
Toxicologist, International Programme on Chemical Safety, Geneva, Switzerland

Dr Edward Smith
Medical Officer, International Programme on Chemical Safety, Geneva, Switzerland

International Agency for Research on Cancer

Dr Henrik Møller
Scientist, Unit of Carcinogen Identification and Evaluation, Lyons, France

WORKING GROUP ON ECOTOXIC EFFECTS LES DIABLERETS, SWITZERLAND, 21–23 SEPTEMBER 1994

Temporary Advisers

Dr Beat Achermann
Air Pollution Control Division, Federal Office of Environment, Forests and Landscape, Berne, Switzerland

Dr Mike Ashmore
Centre for Environmental Technology, Imperial College of Science, Technology and Medicine, London, United Kingdom

Professor J. Nigel B. Bell
Centre for Environmental Technology, Imperial College of Science, Technology and Medicine, London, United Kingdom

Dr Roland Bobbink
Department of Ecology, Section of Environmental Ecology, University of Nijmegen, Netherlands

Dr Thomas Brydges
Canada Centre for Inland Waters, Burlington, Ontario, Canada

Dr Keith Bull
Chairman, UN-ECE Working Group on Effects, Institute of Terrestrial Ecology, Huntingdon, United Kingdom *(Rapporteur)*

Dr Simon J.M. Caporn
Division of Environmental Sciences, Crewe and Alsager Faculty, Manchester Metropolitan University, Crewe, United Kingdom

Dr Ludger van der Eerden
AB-DLO, Centre "De Born", Wageningen, Netherlands

Dr Peter H. Freer-Smith
Principal, Environmental Research, Forestry Commission, Farnham, Surrey, United Kingdom

Dr Jürg Fuhrer
Swiss Federal Research Station for Agricultural Chemistry and Environmental Hygiene, Liebefeld-Berne, Switzerland

Dr Jean-Paul Hettelingh
Deputy Head, Bureau for Environmental Forecasting, National Institute of Public Health and Environmental Protection, Bilthoven, Netherlands

Dr Juha Kämäri
University of Helsinki, Helsinki, Finland

Dr Georg H.M. Krause
Landesanstalt für Immissionsschutz NRW, Essen, Germany

Professor John A. Lee
School of Biological Sciences, University of Manchester, Manchester, United Kingdom

Dr Jan Roelofs
University of Nijmegen, Section Environmental Ecology, Nijmegen, Netherlands

Dr Lena Skärby
Swedish Environmental Research Institute, Gothenburg, Sweden

Dr Robin Wilson
"Talworth", Carperby, Nr Leyburn, North Yorks, United Kingdom *(Chairperson)*

Representatives of other organizations

United Nations Economic Commission for Europe

Dr Radovan Chrast
Environment and Human Settlements Division, Air Pollution Section, Geneva, Switzerland

European Commission

Ms Kathleen Cameron
Detached National Expert, DG X1-B3, Brussels, Belgium

World Health Organization

Regional Office for Europe

Dr Maged Younes
Toxicologist, WHO European Centre for Environment and Health, Bilthoven, Netherlands

Headquarters

Dr Bingheng Chen
Toxicologist, International Programme on Chemical Safety, Geneva, Switzerland

WORKING GROUP ON "CLASSICAL" AIR POLLUTANTS, BILTHOVEN, NETHERLANDS, 11–14 OCTOBER 1994

Temporary Advisers

Dr Ursula Ackermann-Liebrich
Department of Social and Preventive Medicine, University of Basle, Switzerland

Dr Bert Brunekreef
Department of Epidemiology and Public Health, Agricultural University, Wageningen, Netherlands

Dr Douglas W. Dockery
Harvard School of Public Health, Environmental Epidemiology Program, Boston, MA, USA

Dr Lawrence J. Folinsbee
Health Effects Research Laboratory, US Environmental Protection Agency, Research Triangle Park, NC, USA

Dr Judith A. Graham
Environmental Criteria and Assessment Office, US Environmental Protection Agency, Research Triangle Park, NC, USA

Dr Lester D. Grant
Environmental Criteria and Assessment Office, US Environmental Protection Agency, Research Triangle Park, NC, USA

Professor Wieslaw Jedrychowski
Chair of Epidemiology and Preventive Medicine, Jagiellonian University, Cracow, Poland

Dr Klea Katsouyanni
School of Medicine, Department of Hygiene and Epidemiology, University of Athens, Athens, Greece

Dr John Christian Larsen
Institute of Toxicology, National Food Agency, Ministry of Health, Søborg, Denmark

Professor Morton Lippmann
NYU Medical Center, Nelson Institute of Environmental Medicine, Anthony J. Lanza Research Laboratories, Tuxedo, NY, USA

Dr Robert L. Maynard
Air Pollution Section, Department of Health, London, United Kingdom

Dr Peter J.A. Rombout
Laboratory of Pharmacology and Toxicology, National Institute of Public Health and Environmental Protection, Bilthoven, Netherlands

Dr Raimo O. Salonen
National Public Health Institute, Division of Environmental Health, Kuopio, Finland

Dr Katarina Victorin
Institute of Environmental Medicine, Stockholm, Sweden

Dr Robert E. Waller
Department of Health, Health Promotion (Medical) Division, London, United Kingdom

Dr H. Erich Wichmann,
Professor of Labour, Safety and Environmental Medicine, Wuppertal, Germany

Dr Denis Zmirou
Department of Public Health, Epidemiology, Economy of Health and Preventive Medicine, Grenoble Medical School, Grenoble, France

Observers

Dr Peter Strähl
Air Pollution Control Division, Federal Office of Environment, Forests and Landscape, Berne, Switzerland

Dr Martin L. Williams
Science Unit, Air Quality Division, Department of the Environment, London, United Kingdom

Representatives of other organizations

European Commission

Ms Kathleen Cameron
Detached National Expert, DG XI-B3, Brussels, Belgium

World Health Organization

Regional Office for Europe

Dr Michal Krzyzanowski
Epidemiologist, WHO European Centre for Environment and Health, Bilthoven, Netherlands

Dr Maged Younes
Toxicologist, WHO European Centre for Environment and Health, Bilthoven, Netherlands

Headquarters

Dr Bingheng Chen
Toxicologist, International Programme on Chemical Safety, Geneva, Switzerland

WORKING GROUP ON INORGANIC AIR POLLUTANTS DÜSSELDORF, GERMANY, 24–27 OCTOBER 1994

Temporary Advisers

Professor Vladimír Bencko
Institute of Hygiene and Epidemiology, First Medical Faculty, Charles University, Prague, Czech Republic

Professor Alfred Bernard
Industrial Toxicology Unit, Faculty of Medicine, Catholic University of Louvain, Brussels, Belgium

Dr J. Michael Davis
Senior Health Scientist, Environmental Criteria and Assessment Office, US Environmental Protection Agency, Research Triangle Park, NC, USA

Dr Bruce A. Fowler
Director, Toxicology Program, University of Maryland Baltimore County, Baltimore, MD, USA

Professor Lars T. Friberg
Division of Metals and Health, Institute of Environmental Medicine, Karolinska Institute, Stockholm, Sweden

Professor Marek Jakubowski
Scientific Secretary, Nofer's Institute of Occupational Medicine, Lodz, Poland

Professor George Kazantzis
Centre for Environmental Technology, Royal School of Mines, Imperial College of Science, Technology and Medicine, London, United Kingdom

Dr Sverre Langård
Department of Occupational and Environmental Medicine, Telemark Central Hospital, Skien, Norway

Dr F.X. Rolaf van Leeuwen
Toxicology Advisory Centre, National Institute of Public Health and Environmental Protection, Bilthoven, Netherlands

Dr Paul Mushak
PB Associates, Durham, NC, USA

Dr Jesper Bo Nielsen
Department of Environmental Medicine, Odense University, Odense, Denmark

Professor Tor Norseth
National Institute of Occupational Health, Oslo, Norway

Professor Fritz Schweinsberg
Abteilung für Allgemeine Hygiene und Umwelthygiene, Hygiene-Institut der Universität Tübingen, Tübingen, Germany

Professor Gerhard Winneke
Medizinisches Institut für Umwelthygiene an der Heinrich-Heine-Universität Düsseldorf, Abt. Psychophysiologie, Düsseldorf, Germany

Dr Yami Yaffe
Mevasseret Zion, Israel

Representatives of other organizations

European Commission

Ms Kathleen Cameron
Detached National Expert, DG XI-B3, Brussels, Belgium

World Health Organization

Regional Office for Europe

Dr Maged Younes
Toxicologist, WHO European Centre for Environment and Health, Bilthoven, Netherlands

Headquarters

Dr Bingheng Chen
Toxicologist, International Programme on Chemical Safety, Geneva, Switzerland

WORKING GROUP ON CERTAIN INDOOR AIR POLLUTANTS HANOVER, GERMANY, 27–29 MARCH 1995

Temporary Advisers

Dr Steven Bayard
Office of Research and Development, US Environmental Protection Agency, Washington, DC, USA (*Co-rapporteur*)

Professor Jean Bignon
Service de Pneumologie et de Pathologie Professionnelle, Centre Hospitalier Intercommunal, Créteil, France

Dr Manfred Fischer
Institute for Water, Soil and Air Hygiene, Federal Environmental Agency, Berlin, Germany

Dr Francesco Forastiere
Epidemiology Unit, Lazio Region Health Authority, Rome, Italy

Dr Jennifer Jinot
Office of Research and Development, US Environmental Protection Agency, Washington, DC, USA (*Co-rapporteur*)

Dr James McLaughlin
Physics Department, University College Dublin, Belfield, Dublin, Ireland (*Chairperson*)

Dr Hartwig Muhle
Fraunhofer-Institut für Toxikologie und Aerosolforschung, Hanover, Germany (*Vice-chairperson*)

Dr Lucas M. Neas
Environmental Epidemiology Program, Harvard School of Public Health, Boston, MA, USA

Professor Göran Pershagen
Division of Epidemiology, Institute of Environmental Medicine, Karolinska Institute, Stockholm, Sweden

Dr Linda Shuker
Institute for Environment and Health, University of Leicester, Leicester, United Kingdom

Dr Gerard M.H. Swaen
Department of Epidemiology, Faculty of Health Sciences, University of Limburg, Maastricht, Netherlands

Dr H.-Erich Wichmann
Director, Institute of Epidemiology, GSF-Forschungszentrum für Umwelt und Gesundheit GmbH, Oberschleissheim, Germany

World Health Organization

Regional Office for Europe

Dr Michal Krzyzanowski
Epidemiologist, WHO European Centre for Environment and Health, Bilthoven, Netherlands

Dr Maged Younes
Toxicologist, WHO European Centre for Environment and Health, Bilthoven, Netherlands

Headquarters

Dr Bingheng Chen
Toxicologist, International Programme on Chemical Safety, Geneva, Switzerland

International Agency for Research on Cancer

Dr Timo Partanen
Unit of Environmental Cancer Epidemiology, Lyons, France

WORKING GROUP ON PCBS, PCDDS AND PCDFS MAASTRICHT, NETHERLANDS, 8–10 MAY 1995

Temporary Advisers

Professor Ulf G. Ahlborg
Unit of Toxicology, Institute of Environmental Medicine, Karolinska Institute, Stockholm, Sweden (*Rapporteur*)

Dr Hans Beck
Bundesinstitut für Gesundheitlichen Verbraucherschutz und Veterinärmedizin, Berlin, Germany

Dr Martin van den Berg
Associate Professor, Environmental Toxicology, Research Institute of Toxicology, University of Utrecht, Utrecht, Netherlands

Dr Linda S. Birnbaum
Director, Environmental Toxicology Division, Health Effects Research Laboratory, US Environmental Protection Agency, Research Triangle Park, NC, USA

Professor Erik Dybing
Director, Department of Environmental Medicine, National Institute of Public Health, Oslo, Norway (*Chairman*)

Professor Hanspaul Hagenmaier
Institute of Organic Chemistry, University of Tübingen, Tübingen, Germany

Professor Jos C.S. Kleinjans
Department of Health Risk Analysis and Toxicology, University of Limburg, Maastricht, Netherlands

Dr F.X. Rolaf van Leeuwen
Toxicology Advisory Centre, National Institute of Public Health and Environmental Protection, Bilthoven, Netherlands

Professor Christoffer Rappe
Institute of Environmental Chemistry, University of Umeå, Umeå, Sweden

Dr Stephen Safe
Department of Veterinary Physiology and Pharmacology, Texas A & M University, College Station, TX, USA

World Health Organization

Regional Office for Europe

Dr Maged Younes
Manager, Chemical Safety, WHO European Centre for Environment and Health, Bilthoven, Netherlands

WORKING GROUP ON VOLATILE ORGANIC COMPOUNDS BRUSSELS, BELGIUM, 2–6 OCTOBER 1995

Temporary Advisers

James A. Bond
Chemical Industry Institute of Toxicology, Research Triangle Park, NC, USA

Dr Gyula Dura
National Institute of Hygiene, Budapest, Hungary

Professor Victor J. Feron
Senior Scientist, Organization for Applied Scientific Research, Zeist, Netherlands

Professor James J.A. Heffron
Department of Biochemistry, Analytical Biochemistry and Toxicology Laboratory, University College, Cork, Ireland

Mr Paul J.C.M. Janssen
Toxicology Advisory Centre, National Institute of Public Health and Environmental Protection, Bilthoven, Netherlands

Dr John Christian Larsen
Institute of Toxicology, National Food Agency, Søborg, Denmark

Dr Robert L. Maynard
Department of Health, London, United Kingdom (*Vice-chairperson*)

Dr M.E. Meek
Environmental Health Directorate, Bureau of Chemical Hazard, Health Canada, Ottawa, Canada

Dr Lars Mølhave
Institute of Environmental and Occupational Medicine, Århus University, Århus, Denmark

Dr Paul Mushak
PB Associates, Durham, NC, USA (*Rapporteur*)

Dr Antonio Mutti
Laboratory of Industrial Toxicology, University of Parma Medical School, Parma, Italy

Dr Hannu Norppa
Laboratory of Molecular and Cellular Toxicology, Finnish Institute of Occupational Health, Helsinki, Finland

Professor Bernd Seifert
Federal Environmental Agency, Institute for Water, Soil and Air Hygiene, Berlin, Germany (*Chairperson*)

Dr Katarina Victorin
Institute of Environmental Medicine, Karolinska Institute, Stockholm, Sweden

Observers

Dr A. Robert Schnatter
Conservation of Clean Air and Water Europe, Brussels, Belgium

Dr Rob O.F.M. Taalman
European Centre for Ecotoxicology and Toxicology of Chemicals, Brussels, Belgium

Representatives of other organizations

European Commission

Ms Kathleen Cameron
Detached National Expert, DG XI-B3, Brussels, Belgium

World Health Organization

Regional Office for Europe

Dr F.X. Rolaf van Leeuwen
Manager, Chemical Safety, WHO European Centre for Environment and Health, Bilthoven, Netherlands

Headquarters

Dr Maged Younes
Chief, Assessment of Risk and Methodologies, Programme for the Promotion of Chemical Safety, Geneva, Switzerland

International Agency for Research on Cancer

Dr Timo J. Partanen
Visiting Scientist, Environmental Cancer Epidemiology Unit, Lyons, France

WORKING GROUP ON "CLASSICAL" AIR POLLUTANTS BRUSSELS, BELGIUM, 6–7 JUNE 1996

Temporary Advisers

Professor Ursula Ackermann-Liebrich
Department of Social and Preventive Medicine, University of Basle, Basle, Switzerland

Professor Bert Brunekreef
Director, Department of Epidemiology and Public Health, Agricultural University, Wageningen, Netherlands

Dr Norbert Englert
Department of Air Hygiene, Federal Environmental Agency, Institute for Water, Soil and Air Hygiene, Berlin, Germany

Dr Judith A. Graham
Acting Associate Director for Health, National Exposure Research Laboratory, US Environmental Protection Agency, Research Triangle Park, NC, USA

Dr Lester D. Grant
Director, National Center for Environmental Assessment, US Environmental Protection Agency, Research Triangle Park, NC, USA

Dr Klea Katsouyanni
Department of Hygiene and Epidemiology, University of Athens School of Medicine, Athens, Greece

Dr Robert L. Maynard
Air Pollution Section, Department of Health, London, United Kingdom

Dr Peter J.A. Rombout
Laboratory of Pharmacology and Toxicology, National Institute of Public Health and Environmental Protection, Bilthoven, Netherlands

Observers

Dr John F. Gamble
Occupational Health and Epidemiology Division, EXXON Biomedical Sciences Inc., East Milestone, NJ, USA

Dr H. Visser
Statistician, KEMA Environmental Services, Arnhem, Netherlands

Representatives of other organizations

European Commission

Ms Lynne Edwards
DG XI-D3, Brussels, Belgium

Mr Pierre Hecq
DG XI-D3, Brussels, Belgium

Dr Canice Nolan
DG XII-D1, Brussels, Belgium

World Health Organization

Regional Office for Europe

Dr Michal Krzyzanowski
Epidemiologist, WHO European Centre for Environment and Health, Bilthoven, Netherlands

Dr F.X. Rolaf van Leeuwen
Manager, Chemical Safety, WHO European Centre for Environment and Health, Bilthoven, Netherlands

Headquarters

Dr Maged Younes
Chief, Assessment of Risk and Methodologies, Programme for the Promotion of Chemical Safety, Geneva, Switzerland

FINAL CONSULTATION ON UPDATING AND REVISION OF THE AIR QUALITY GUIDELINES FOR EUROPE BILTHOVEN, NETHERLANDS, 28–31 OCTOBER 1996

Temporary Advisers

Dr Beat Achermann
Air Pollution Control Division, Federal Office of Environment, Forests and Landscape, Berne, Switzerland

Dr Mike Ashmore
Centre for Environmental Technology, Imperial College of Science, Technology and Medicine, London, United Kingdom

Professor Alfred Bernard
Industrial Toxicology Unit, Faculty of Medicine, Catholic University of Louvain, Brussels, Belgium

Dr Roland Bobbink
Department of Ecology, Section of Environmental Biology, University of Nijmegen, Nijmegen, Netherlands

Professor Bert Brunekreef
Department of Epidemiology and Public Health, Agricultural University, Wageningen, Netherlands

Ms Kathleen Cameron
Head, Chemicals and Health Branch, Department of the Environment, London, United Kingdom

Professor Erik Dybing
Director, National Institute of Public Health, Department of Environmental Medicine, Oslo, Norway *(Chairperson)*

Dr Ludger van der Eerden
AB-DLO, Centre "De Born", Wageningen, Netherlands

Professor Victor J. Feron
Senior Scientist, Netherlands' Organization for Applied Scientific Research, Zeist, Netherlands

Dr Judith A. Graham
Acting Associate Director for Health, National Exposure Research Laboratory, US Environmental Protection Agency, Research Triangle Park, NC, USA

Dr Lester Grant
Director, National Center for Environmental Assessment, US Environmental Protection Agency, Research Triangle Park, NC, USA

Dr Jean-Paul Hettelingh
Deputy Head, Bureau for Environmental Assessment, National Institute of Public Health and Environment, Bilthoven, Netherlands

Professor Marek Jakubowski
Scientific Secretary, Nofer's Institute of Occupational Medicine, Lodz, Poland

Professor Matti Jantunen
KTL Environmental Health, EXPOLIS, Helsinki, Finland

Dr John Christian Larsen
Institute of Toxicology, National Food Agency of Denmark, Ministry of Health, Søborg, Denmark

Professor Morton Lippmann
NYU Medical Center, Nelson Institute of Environmental Medicine, Anthony J. Lanza Research Laboratories, Tuxedo, NY, USA

Professor Marco Maroni
Director, International Centre for Pesticide Safety, Busto Garolfo, Italy

Dr Robert L. Maynard
Head, Air Quality Science Unit, Department of Health, London, United Kingdom *(Rapporteur)*

Dr James P. McLaughlin
Radon Research Group, Department of Experimental Physics, University College Dublin, Dublin, Ireland

Dr M. Elizabeth Meek
Head, Priorities Substances Section, Environmental Health Centre, Health Canada, Ottawa, Ontario, Canada

Dr Lars Mølhave
Institute of Occupational and Environmental Medicine, Århus University, Århus, Denmark

Dr Peter J.A. Rombout
Toxicologist, Laboratory for Health Effect Research, National Institute of Public Health and Environment, Bilthoven, Netherlands

Professor Bernd Seifert
Director, Department of Air Hygiene, Federal Environmental Agency, Institute for Water, Soil and Air Hygiene, Berlin, Germany

Professor H. Erich Wichmann
Director, GSF Institute of Epidemiology, Research Centre for Environment and Health, Oberschleissheim, Germany

Dr Martin L. Williams
AEQ Division, Department of the Environment, London, United Kingdom

Professor Gerhard Winneke
Medical Institute for Environmental Hygiene, Heinrich-Heine University, Düsseldorf, Germany

Observers

Dr Masanori Ando
Director, Division of Environmental Chemistry, National Institute of Health Sciences, Ministry of Health and Welfare, Tokyo, Japan

Mr Soichiro Isobe
Deputy Director for Chemicals Safety Management, Office of Environmental Chemicals Safety, Environmental Health Bureau, Ministry of Health and Welfare, Tokyo, Japan

Representatives of other organizations

European Commission

Ms Lynne Edwards
DG XI, D-3, Brussels, Belgium

World Health Organization

Regional Office for Europe

Dr Michal Krzyzanowski
Epidemiologist, European Centre for Environment and Health, Bilthoven, Netherlands

Dr F.X. Rolaf van Leeuwen
Manager, Chemical Safety, European Centre for Environment and Health, Bilthoven, Netherlands

Headquarters

Dr Bingheng Chen
Toxicologist, International Programme on Chemical Safety, Geneva, Switzerland

Dr Maged Younes,
Chief, Assessment of Risk and Methodologies, Programme for the Promotion of Chemical Safety, Geneva, Switzerland

International Agency for Research on Cancer

Dr Douglas B. McGregor,
Scientist, Unit of Carcinogen Identification and Evaluation, Lyons, France

WORKING GROUP ON GUIDANCE FOR SETTING AIR QUALITY STANDARDS
BARCELONA, SPAIN, 12–14 MAY 1997

Temporary Advisers

Dr Beat Achermann
Air Pollution Control Division, Federal Office of Environment, Forests and Landscape, Berne, Switzerland *(Rapporteur)*

Dr Alena Bartonova
Norwegian Institute for Air Research (NILU), Kjeller, Norway

Mr Xavier Guinart
Environmental Department, Air Monitoring and Control Service, Generalitat de Catalunya, Barcelona, Spain

Dr Oriol Puig I Godes
Environmental Department, General Subdirectorate of Air Quality and Meteorology, Generalitat de Catalunya, Barcelona, Spain

Mr Harvey Richmond
Risk and Exposure Assessment, Air Quality Planning and Standards, US Environmental Protection Agency, Research Triangle Park, NC, USA

Professor Bernd Seifert
Director, Department of Air Hygiene, Federal Environmental Agency, Institute for Water, Soil and Air Hygiene, Berlin, Germany *(Chairperson)*

Dr Éva Vaskövi
Department of Air Hygiene, Johan Béla National Institute of Public Health, Budapest, Hungary

Dr Martin L. Williams
AEQ Division, Department of the Environment, London, United Kingdom

Observers

Dr Merce Aceves
Head, Environmental Laboratory, Generalitat de Catalunya, Barcelona, Spain

Professor Guillem Massague
Head, Air Monitoring and Control Service, Environmental Department, Generalitat de Catalunya, Barcelona, Spain

Dr Joseph Eduard Mata
General Subdirector of Health Protection, Department of Public Health and Social Security, Generalitat de Catalunya, Barcelona, Spain

World Health Organization

Regional Office for Europe

Dr Michal Krzyzanowski
Environmental Epidemiologist, WHO European Centre for Environment and Health, Bilthoven, Netherlands

Dr F.X. Rolaf van Leeuwen
Manager, Chemical Safety, WHO European Centre for Environment and Health, Bilthoven, Netherlands

Headquarters

Dr Dieter Schwela
Urban Environmental Health, Division of Operational Support in Environmental Health